# THE ANANDROUS JOURNEY

"life on her own without husband,"
also a botanical term familiar to Merrill's studies

*Harriet Bell Merrill*

## Revealing Letters to a Mentor

by Merrillyn Leigh Hartridge

Copyright © 1997 Merrillyn Leigh Hartridge

First Edition

All rights reserved. Reproduction in whole or in part of any
portion in any form without permission of publisher is prohibited.

Cover art by M. L. Hartridge

Printed in the United States of America by
Palmer Publications, Inc.
PO Box 296
Amherst, WI 54406

**Library of Congress Cataloging-in-Publication Data**

Hartridge, Merrillyn Leigh
    The anandrous journey: revealing letters to a
mentor / by Merrillyn Leigh Hartridge. — 1st ed.
      p.  cm.
    ISBN: 0-942495-66-7. — ISBN: 0-942495-65-9 (pbk.)
    1. Merrill, Harriet Bell, 1863—Correspondence.  2. Women
zoologists—Correspondence.  3. Merrill, Harriet Bell, 1863—
Journeys—South America.  4. Zoology—South America.  5. Zoological
specimens—South America—Collection and preservation.  6. South
America—Description and travel.     I. Title.
QL31.M553A4 1997
590'.92—dc21
[B]
                                                97-7950
                                                  CIP

Library of Congress No.: 97-7950

Palmer Publications, Inc.          Amherst Press

# Dedication

To my family
the past, present and future generations

ESTADOS UNIDOS
DE
VENEZUELA
TARJETA POSTAL
UNIÓN POSTAL UNIVERSAL.
(LADO RESERVADO

Curaçao. Scharloo y Pietermaai

Gathmann Hnos., Carácas. 05520.
Arreo de burros.
SALUDO DE VENEZUELA.

# Contents

Acknowledgments ................................................................... vii
Credentials of Harriet Bell Merrill .................................................. ix
Profile of Harriet Bell Merrill ....................................................... xi
Introduction to The Anandrous Journey ......................................... xiii

**Part I**
**A Biographical Narrative Based on Harriet Bell Merrill's Letters and Career**
    **as Research Assistant to Dean Edward Asahel Birge** ....................... 1
    Kappa Alpha Theta ............................................................... 7

**Part II**
**Journey Through South America** ................................................. 23
    Journey Through South America by Enterprising Milwaukee Woman ........ 24
    Merrill Really Rolls to Rio Sailing Out of New York on *S. S. Byron* ........... 25
    Arrival in South America ........................................................ 26
    Flora and Fauna and the Botanical Gardens of Rio de Janeiro ............... 30
    Garden of Decadence ............................................................ 32
    Trains—The Unique Cog Rail ................................................... 34
    Boots Are Objects of Curiosity .................................................. 36
    Officials for the World's Fair in St. Louis ...................................... 37
    South American Oil Industry ................................................... 38
    Rubber Trees and Plantations .................................................. 39
    A Gem—Set in the Prongs of the Amazon ................................... 40
    Culture of the Rain Forest Dwellers ........................................... 41
    The Big Fishing Expedition ..................................................... 43
    Steamer Trunks and Intriguing Passengers .................................. 44
    *R.M.S. Magdalena* .............................................................. 45
    Currency "Control" .............................................................. 46
    Nightmare Trek to Iguazú Falls ................................................ 50
    Trek to Iguazú Falls ............................................................. 51
    Light in the Rain Forest ......................................................... 52
    The Brown Amazon and Tributaries ........................................... 53
    On to Asunción Unaccompanied by Men ..................................... 54
    History of Paraguay ............................................................. 56
    A Jesuit Stronghold ............................................................. 56
    The Church and the Missionaries .............................................. 57
    Career of Younger López ....................................................... 58
    López Plunged Country Into War Venezuelan Conflict ...................... 59
    The Beef and Leather Industry ................................................ 61
    The Guest of Honor ............................................................. 64
    Latino Beauties and Exuberant Equestrians at Their Exhilarating Best .... 66
    Observations on Homes and Social Customs ................................ 67
    Life in the Consulates .......................................................... 70

    The Dumont Coffee Cartel................................................. 73
    The Dumont Family......................................................... 74
    Cultural Customs and Contrasts in South America............... 74
    Stopping the Streetcars.................................................... 76
    Selection of Food............................................................ 77
    Two Distinctively National Dishes..................................... 78
    Fiesta Days in Rural Areas................................................ 79
    Water, Water Everywhere—What to Drink?....................... 81
    Dr. Cruz—Medical Leader Extraordinaire........................... 82
    The Continent of Great Potential...................................... 86
    Expatriots Hide Specious Backgrounds.............................. 87
    Debtors Many in Paraguay............................................... 88
    Factories are Few............................................................ 89
    Opportunities for Investors.............................................. 89
    Good Land Comes High................................................... 91
    The "Spider" Women of South America............................ 92
    Clothing Customs........................................................... 94
    The Fashionable Ciudadanos............................................ 97
    Customs in Schools......................................................... 98
    Elementary Level Studies................................................. 98
    The Modern Woman of South America............................. 100
    The Foreign Women Have Clubs...................................... 102

**Part III**
**Back on Terra Familiar**............................................................ 103
    Harriet Merrill's Return to Appointed Tasks...................... 105
    1904—A Historic Year for the Capitol of Wisconsin.......... 107
    The Intriguing Legacy From Young America..................... 108
    All American-Germans................................................... 115
    A Truly Gemütlich Gathering.......................................... 118

**Part IV**
**The Birge—Merrill Correspondence**..................................... 125
    Using Some of the Last Good Light................................. 160
    Post Death Correspondence........................................... 167

**Part V**
**Discovery of the E. A. Birge and H. B. Merrill Letters**........... 173
    The First Inquiry to Dr. Frey........................................... 178
    Charts of H. B. Merrill's Scientific Research..................... 186

Chapter notes
    1. Reflections of Birge by Colleagues............................... 191
    2. Chancey Juday—A Birge Associate............................... 192
    3. Academic Criteria—Early 1900s.................................. 193
    4. Birge Earns Ph.D......................................................... 194
    5. Student Housing and Greek Letter Societies................ 194
    6. University Politics...................................................... 196
    7. Birge's Private Life..................................................... 198
    8. The Allen Survey........................................................ 198
Index............................................................................................ 201
About the Author....................................................................... 213

# Acknowledgments

I begin my thanks to my dear husband, Dr. Theodore Livingston Hartridge (University of Pennsylvania Medical School). As an E.N.T. surgeon, he practiced and taught in Madison, Wisconsin, and was acquainted with certain University of Wisconsin faculty members mentioned by H. B. Merrill, as well as those she referred to who were from well-known Milwaukee families, where he was born. Ted Hartridge has a keen memory and fondness for history and always has an appreciation for my interests. I am grateful to my thoughtful, dear daughter, Lynn Ann (Nan) Casper, whose meticulous research as an artist historian is widely acclaimed not only for the accuracy of her architectural illustrations, but also her cityscapes and maritime scenes. Between her home, family and business commitments to fine art collectors, she has been my most knowledgeable critic.

Thanks to Susan Otto and Judy Turner in the photographic index section of the Milwaukee Public Museum and Carter Lupton, curator in the anthropology division who provided me with lists of acquisitions from the Merrill estate numbering hundreds of zoological, botanical, biological and ethnological items that are catalogued there.

I deeply appreciate the foresight of Mrs. Nathan E. Merrill who kept Harriet Bell Merrill's photographs which were taken in the early 1900s and considered quite rare, particularly of subjects captured in South America by a woman from North America. They, along with family memorabilia, are a valued legacy to the resurrection of Merrill and her era.

My thanks to Dr. Stanley I. Dodson of the University of Wisconsin Department of Zoology who referred me to Emeritus Professor Arthur D. Hasler of the University of Wisconsin Center for Limnology. A colleague of E. A. Birge, Hasler suggested that paleolimnologist Professor David G. Frey of the University of Indiana was probably the most knowledgeable of Merrill's study. I have devoted a section on Professor Frey which describes the discovery and location of the Birge—Merrill papers in Part V, page 173. Thanks also to Professor John Bennet, Classics Department, University of Wisconsin, Madison, and Milton R. Varsos, psychologist of the Wisconsin schools systems, both well versed in literature and linguistics. Appreciation also is given to Carol Butts, historian archivist, Lawrence University, Appleton, Wisconsin, and Dr. John Dallman, curator for the University of Wisconsin Zoological Museum, Mr. Bernard Schermetzler, University of Wisconsin Memorial Library Archives, and Susan Ravdin, special collections division of Bowdoin College Library, Brunswick, Maine.

A thank you to Linda Holthaus and Anna Marie Beckell of the University of Wisconsin Limnology Center. Anna Marie recognized from my research that Merrill deserved recognition and with the consensus of the center, was instrumental in arranging for the dedication of a building at Trout Lake Research Station in the name of Harriet Bell Merrill. My thanks to Faith Miracle, editor of the *Wisconsin Academy of Sciences, Arts and Letters.* Her very name gave meaning to my hope of restoring the memory of the Academy's first woman to serve as vice president on the Wisconsin Academy council. I was heartened by responses from Dr. J. W. Reid of the Smithsonian Museum of Natural History, Washington, DC, and Dr. Dorothy Berner of Temple University, Philadelphia, who are continuing in Merrill's discipline on the study of Cladocera. It was through them, I learned that much of Merrill's copepod material is still in existence, held for further research at the Smithsonian. I am grateful to Dr. Allen Walker Read, New York, Rhodes Scholar, Professor Emeritus of English at Columbia University who holds numerous additional academic honors and was a colleague of H. L. Mencken, collaborating on lexicography texts. I am also indebted to the State Historical Society of Wisconsin's visual material and genealogical divisions; to my beloved mother, Florence Hess Leigh, whose memory of the past and her Aunt Hattie Bell Merrill, remained sharp to the end of her 95 years; to University of Illinois Archives, Urbana-Champaign and Archive Assistant Robert T. Chapel who turned over biographical material to me on Charles Zeleny and H. B. Ward in whose departments Merrill taught and worked toward her Ph.D.

I was fascinated by the *Wisconsin Academy Review* cover with caricatures of notable members of the Society marking its 125th year, particularly the sketches of Harriet Bell Merrill depicting her discipline, all rendered by Warrington Colescott and John Wilde; and to the music of Vladimir Cosma, Strunz and Farah (particularly *Amazanas*) and the bird sounds of the rain forests which were mood motivators accompanying the "typewriter castanets," as I wrote of South America.

I am sincerely indebted to the many positive responses and encouragement from professional associates who read the initial publication of the *Wisconsin Academy Review 125th Year Special Issue* on H. B. Merrill and suggested that her life would make a fascinating historical biography. I am thankful to Reverend John R. Thomas, Dr. Louis Warrick and long-time friends with whom I have shared life's many reversals and crises—theirs and mine—and who have continued to encourage my projects.

I thank Mary Edith Arnold, archivist Kappa Alpha Theta National for the history of the charter of the Psi chapter, established May 29, 1890, in Madison, Wisconsin.

# Credentials of Harriet Bell Merrill

In 1900 Harriet Bell Merrill joined the faculty at the University of Wisconsin as an assistant professor reporting to Dr. Edward Asahel Birge, chair of the university's Department of Zoology. Her experience as head of the science department at Milwaukee-Downer College and the years of teaching in Milwaukee high schools had prepared her for this role. She found that her studies in microbiology required more extensive monitoring of a Cladocera in the family Macrothricidae, a study that Birge himself had begun, but found little time to research. She had written a monograph on the systematics and anatomy of the genus *Daphnia*; at the time it was published, she could not have known that it would become a lifelong pursuit.

Harriet Bell Merrill was the first woman to hold the position of vice president on the council of Sciences, Arts and Letters of the Wisconsin Academy. She was elected in 1896, and in 1899 she succeeded Frederick Jackson Turner as vice president of Letters and Sciences, a position she held until her departure for South America in 1902. Harriet Bell Merrill graduated from the University of Wisconsin *summa cum laude* in 1890 with a B.S. degree, and was a member of Laurea Honorary Society. She did graduate work at Cornell, Chicago, and Wisconsin universities; and received an M.S. degree from Wisconsin in 1893. She was head of the science department at Milwaukee-Downer College (1897-99), where she taught courses in general chemistry, organic chemistry, general biology, zoology, vertebrate and invertebrate anatomy, physiology, and psychology. She was director of the physiology and biology departments at Milwaukee's East and South high schools (1890-99); was elected honorary fellow of the University of Chicago (1894-99); did research work during the summer of 1893 at Woods Hole Biological Laboratories (1898-1900); lectured alternately at Cornell, Chicago, and Wisconsin universities in departments of zoology, economics, biology, and botany. She was elected to the Wisconsin chapter of the Natural History Society on October 31, 1901.

During the period of 1902-03, Merrill traveled throughout South America on a research tour, retracing her contacts again in 1907-09. She returned to the University of Wisconsin as an assistant professor of zoology, with requests to lecture on her experiences at the universities where she was associated and at the Milwaukee Public Museum. At the time of her death in 1915, she was enrolled in the Ph.D. program at the University of Chicago.

In 1990 a building at Trout Lake Research Station at Boulder Junction was named for H. B. Merrill by the University of Wisconsin-Madison Center for Limnology.

*This photo was taken in 1890 when Harriet Bell Merrill graduated from the University of Wisconsin.*

—*Photo courtesy of the State Historical Society of Wisconsin.*

Baixo-Amazonas.  Santarem — Rio Tapajós — Pará. (Praia)

UNION POSTALE UNIVERSELLE
BRAZIL
CARTE POSTALE - BILHETE POSTAL.

Espaço reservado á correspondencia.  Espaço reservado á direcção.

Dear Dr. Birge,
I am as misplaced as
a wood violet in this
Mesolithic Eden but
I am getting some good
live specimens for you.
Wish you were here
H.B.M.

Dr. E. A. Birge
The University of Wisconsin
U.S.A.

# Profile of Harriet Bell Merrill

## Merrill's Life a Journey Without a Husband

Harriet Bell Merrill's letters and articles, follow her career from her days as a student, teacher, assistant professor and finally, a researcher for Dr. Edward Asahel Birge. Though she considered him a mentor, the correspondence from almost one hundred years ago reveals that she was a dedicated confidant and friend. Through more than 27 years of her association with Birge and the departments of Zoology and Limnology, she worked with men in the sciences. Merrill followed the dictum observed by most women of her time in that it was more prudent to carry out academic procedures organized by the men in coeducational institutions than for women to attempt to administer their own policies.

The letters indicate that some colleagues and family members felt the adverse living conditions she had to contend with in South America and while at the University of Illinois, contributed to her deteriorating health. They also assumed that she harbored a blind dedication to her mentor and his desire for her to continue the field of research that he considered a priority in his discipline but depended upon others to carry out. Because women were considered physically incapable of withstanding the rigors of hazardous field research, the fact that Merrill attempted and accomplished two lengthy journeys to foreign territories on her own, was notable in the 1900s.

When Hattie Bell emerged from her North American "chrysalis" and spread her wings south of the equator, she described the adventure as being, "akin to paradise." Yet none of her later letters indicated any desire to alter her intent to stay the course where she had begun. She only hoped to encourage the next generation to experience the symbiotic disposition to the sciences as she had. One can almost hear her justification of her decision to return to familiar territory by saying, "Paradise is exhilarating for awhile, but my expectations of beauty have been amply served by not meandering too long in Elysian fields!"

As Frey mentioned, "Merrill was the Birge assistant who traveled further afield than others during her time." She was cautious not to complain about the situations encountered, to her mentor and certainly not to her brothers who were overly concerned because of her cardiac condition. Her life's goal was to complete her Ph.D. and ultimately head a woman's college. Students had been her family, and her abiding interest in the sciences sustained her to the end of her full life of only 52 years.

*Rendena Street, a residential area of Buenos Aires.*
—*H. B. Merrill photo*

# Introduction to The Anandrous Journey

When writing a biographical narrative based upon historical fact, it helps to be a sleuth with a propelling need for continuing research. In the case of woman scientist Harriet Bell Merrill, it has taken me nearly three decades of investigating the clues she left behind before disclosing the story of her eventful life. This narrative is written based upon fact from correspondence and information from the period in which the events took place. Even historians of note make arbitrary choices as to how much of their research is intrinsic to the entire account. No matter how far afield the pursuit, there will inevitably be some document, some person, some scrap of evidence yet to be discovered. The Merrill biographical summary I wrote was first published by *The Wisconsin Academy of Sciences, Arts and Letters* for their 125th year special issue. It was a challenge to completely reconstruct the manuscript to include a broader readership while retaining certain essential scientific data and verification on E. A. Birge. To a degree, it reminds me of the process of painting a portrait or a landscape. At a certain point in the work, a decision has to be made as to whether the addition of one more line or brush stroke will give any greater delineation to the subject.

❧

Harriet Bell Merrill was born in Stevens Point, Wisconsin, February 6, 1863. Her father, Samuel Merrill, left the family mill in Brunswick, Maine, when the lumber business began to boom in Wisconsin. He settled in Portage County in 1858 and married Anna Comstock Emmons, a school teacher from Duchess County, New York. Harriet had two brothers, Nathaniel Emmons Merrill, a patent attorney in Chicago and Roger L. Merrill, Wisconsin agent for Fidelity and Deposit Company of Maryland, a national insurance and bond company (Majestic Building, Milwaukee, Wisconsin).

During the Civil War, the year 1863 marked an attempt to integrate female students into the University of Wisconsin. It was also the year that National Conscription had been enacted by President Lincoln, causing a substantial reduction in male student enrollment in institutions of higher learning.

Professor Charles H. Allen, in March of 1863, was appointed principal of the Normal School which permitted women to take up the slack. Eventually they were integrated into classes with the men. By October of the same year, there was dissension from various sectors on campus opposed to accepting Allen's Normal School into their University. Dean Sterling declared, "The fear is that the admission of females to the privileges of the University will result in letting down the standard of culture if not destroying its character as a college."

For the most part it was decided that female students had a positive effect on the decorum of male students. "For all their physical weakness, they are more punctual and orderly by comparison." Finally, on June 16, 1908, the Board of Regents declared the resolution that "Men and Women shall be equally entitled to membership in all classes at the university." It is interesting to note that by 1890, Merrill had graduated from the University of Wisconsin, received an M.S. degree in 1893 and became a researcher for Birge as assistant professor of zoology (see Credentials of Harriet Bell Merrill page ix).

A fourth generation Madisonian, I grew up in a large family with mostly adults. When the men got together, the talk was about politics, the wars they and their fathers fought—their youthful days when they explored in the Yukon or Australia, or their confrontations with native tribes on the Western Plains. Conversation among the women often turned to conflicting arguments about the virtues of married life. The opposition, comprised of spinster teachers, nurses and others with the limited occupations available to single, educated women, thought the Women's Suffrage movement had not altered many of the gender-biased restrictions in the workplace. The legislation intended to protect women was considered by most to be based on an offensive "separate place" doctrine, restricting their hopes for advancement in the new century. Even with ratification of the 19th Amendment (August 18, 1920), which allowed them the vote, women were still unsure of their place in what they observed was a man's world. My mother's sister caused quite a furor in her family when she accompanied her husband to the Philippine Islands where the two doctors researched dietetics among the Igorot and Moro tribes in the 1920s. She also had children. This would have been highly unlikely for her Aunt Hattie to have realized in the 1890s when marriage and a family obviated plans for any other career for women.

Harriet Bell Merrill, my great-aunt on my maternal grandmother's side, died several years before I was born, but the stories I heard about her heightened my curiosity. My parents, John and Florence Leigh, named me Merrill after her, adding the "lyn" to feminize it. She appeared sweet and demure in her photos and was described by the family as "fair and petite, standing barely five feet tall from her heavy man's boots to the top of her light brown pompadour." My mother said that Merrill's small frame was often encumbered with her camera and scientific equipment when she tromped about Madison lakes and around the campus.

Outside of her research group, few persons had any real comprehension of the importance of the family Macrothricidae, a field of study that the head of the zoology department, Dr. Edward Asahel Birge at the University of Wisconsin, had begun and found little time to further, particularly when he became acting president upon the death of Dr. Allen. Merrill had written a monograph on *The Systematics and Anatomy of the Genus Daphnia,* which was the basis for her M.S. degree, and it was this study that Dr. Birge encouraged her to continue.

During her entire academic career, Merrill matriculated in a predominantly male academic environment. How she prevailed in the fairly restrictive milieu of a midwestern university at the turn of the 19th century is evident in her letters which give straightforward accounts of the social, political and sexual mores of her time. It wasn't until 1986 that an astounding discovery of a large cache of Birge—Merrill letters were uncovered by a limnologist, Dr. E. S. Deevey at the University of Florida. Upon retirement he had planned to place them in the archives of Yale University. The story of their interception and my request to bring Harriet Bell Merrill's letters home to Wisconsin is told in the Discovery of the E. A. Birge and H. B. Merrill Letters, Part V, page 173. They reveal her professional association of 27 years with Dr. Edward Asahel Birge who was dean of the zoology department where she was an assistant professor. Although Merrill had been head of the Milwaukee-Downer Science Department, was elected Honorary Fellow at the University of Chicago where she taught, lectured at Cornell and researched at Woods Hole Biological Laboratories where few women had been admitted (see Credentials of Harriet Bell Merrill, page ix), she still had to cautiously guard her hard-won place.

Men on the University of Wisconsin faculty had an opinion commensurate with other teaching institutions at the time—that women were subordinate to men in the sciences. Some male scholars who were shown a list of H. B. Merrill's achievements were markedly impressed. When they learned that H. B. was a woman, they justified that her interest in continuing research in Birge's discipline was less for the possibility of the discovery of new species, but more a blind devotion to her benefactor. If earlier letters implied this, her later correspondence belied the assumption. Merrill did express to women friends, "When you have finally earned your place, the hardest part will be not to appear too aggressively assertive while firmly claiming your territory!"

While Merrill assumed a didactic approach in expressing professional opinions to her male colleagues, she apparently provided Birge with information through her connections beyond the "Hill," (the general area of university campus with the major building, Bascom Hall, at the top of the hill) with more diverse views than those divulged by his male associates. Their correspondence indicates that he was interested in her sagacious insights through her associations with political factions in the university's town-gown circles and particularly her viewpoints on campus politics. There is almost a sisterly quality—even a mentor role reversal in her letters to Birge at the end of her life.

In 1902 at age 39, Merrill seemed to experience a mid-life self-appraisal that unexpectedly prompted her to demonstrate an independent attitude in a man's world. In fact, as my own life had taken on many dimensions that demanded family time and with other jobs related to writing, my Merrill biography went on stand-by. Then, in 1961, at the death of Mrs. Nathaniel Merrill (Harriet's sister-in-law), personal papers and several mementoes were handed on to me from her estate. Among them were coin silver spoons and fork from the American Colonial period, engraved with the name Emmons (see letter, page 124) and a porcelain tea set (with some missing cups) from England's 18th century commercial trade with Canton, China. The most intriguing legacy was a box labeled simply "HB Merrill Papers." When I examined the contents they revealed much more about Great-aunt Hattie's life as a teacher, particularly her impressions of South America. The Daughters of the American Revolution papers were essential for family history, and transcripts identified her "place" at the University of Wisconsin. Her writing was primarily in papers published in an 1893 copy of the *Transactions of the Wisconsin Academy of Sciences, Arts and Letters* and a few letters to family and friends. But what has been considered a unique find among the items are dozens of sepia-toned photographs taken by Merrill during the turn-of-the-century era. Most of the pictures in the collection are rare views of South America. Directions written on the reverse side of the photos indicated that they had originally been sent along with the articles she wrote for publication in Milwaukee newspapers.

Merrill saw contrasts in cultures between the wealthy Latinos who owned vast estancias, the primitive tribes in rain forests, the order and protocol exhibited by the British and the industrial enterprises of the German immigrants. She traveled by steamboat, cog rail and on horseback into areas where few white women, if any, had ventured. She hazarded the bubonic plague, malaria and cholera everywhere she set foot. Knowing the risks she might encounter, Harriet Merrill still felt a release as liberating as "loosing the constraints of corset stays and wearing a shift," as she left the staid midwest campus and sailed to South America.

Her brothers were opposed to her going alone to what they termed a "hostile country," and although research was the primary motive in selecting destinations for her travels, she had always wanted to see South America. She was so determined to go that other than letters of introduction, which assured her accommodations during her stay at certain consulates and universities while abroad, she was virtually on her own from ship to shore. In a mostly Latin society where single women were expected to be accompanied by a male member of their family or at least a dueña, Hattie Bell Merrill was a conundrum. The manner in which she dealt with this dilemma is evident in her subtle but witty exchanges that make for fascinating reading.

She ultimately made two journeys; the first in 1902-1903 to learn about the country and certain species and then in 1907-1909 for in-depth scientific study. The contrast between the notes from Harriet Merrill's trips are striking, in that, on the second trip she was working at the species level, particularly on Chydoridae and Macrothricidae. Her identifications enabled the records to be used in assessing the cladoceran fauna of South America. In any given region except Antarctica, 30 to 50 species of Cladocera are found in freshwaters. The species that Merrill was pursuing were in the genus *Bunops* which are uncommon and difficult to isolate. Dr. Frey had found them only a couple of times. The importance of these forms of life (less than 1 mm in length) is their relationship to the ecology of the earth's bodies of water.

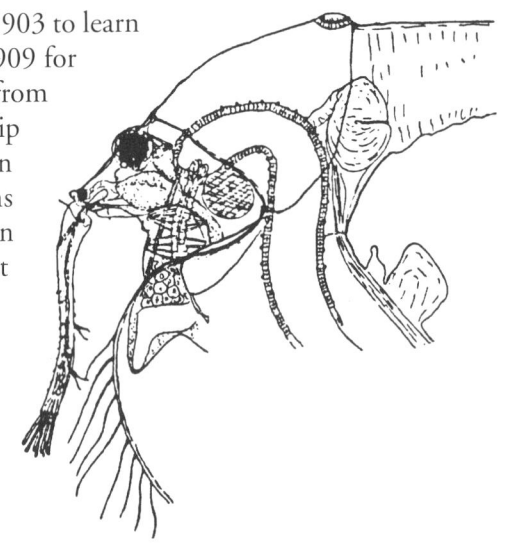

Bunops scutifrons. *Head. X100. a lens-like body. Drawing by H. B. Merrill.* Wisconsin Academy Transactions, *Volume 9, Part II, page 342. 1892-93.*

The varying conditions of gathering specimens en route created problems in keeping aquatic material preserved for chemical analysis and histological examination. Often a day's work might be wasted because of the lack of hydrobiological laboratory facilities in the remote locations. The inconvenience did not seem to deter Merrill. She assiduously collected—on several expeditions—over 700 samples of various genera, marking the location and species in 15 field notebooks. Her work, particularly on *chydorids* and *macrothricids*, resulted in discovery of a species from Brazil named after her—*Diaptomus merrilli* by Dr. Stillman Wright (who received his Ph.D. from Chancey Juday in 1928). He, along with Dr. Birge, Professor Arthur Davis Hasler and Chancey Juday, developed the limnology program at the University of Wisconsin. Their work on these forms of life, essential to the condition of the earth's bodies of water and effects on related ecological factors, continues to be researched. Some of her material remains extant to this day—according to the Smithsonian Institute.

It was Merrill's unflagging pursuit of her particular discipline that noted paleolimnologist Dr. David Frey proclaimed a great contribution to those continuing in her field of science.

Around the turn of the century, there were exciting new concepts in the study of plankton communities in bodies of water. In biological terms, the interlocking network of physical and chemical processes were determined by the migration of species of zooplankton. "The distribution and balance of these animals determines the life of the lakes—ultimately the oceans."[1] In addition, Merrill brought or sent back hundreds of ethnological, anthropological, botanical and zoological specimens which were acquisitioned by the Milwaukee Public Museum where they are catalogued.

Because of the precarious transportation facilities and her unpredictable schedule, Merrill took notes, on the go, reporter-form. If a typewriter were available she used it, sending off completed articles to the Milwaukee newspapers that published them in serial format during the 1902-1903 years.

While en route, it was impossible for Merrill to predict exact dates of the collection of her mail and photographic film or when it would reach the United States. What is important is the content of her articles and letters. With candidly perceptive observations, they give us a glimpse into an absorbing, private life that spanned the period between the post-Civil War in the United States to World War I. The comparisons between the people, places and happenings of that time, may seem extreme in contrast to the present. Then again, many aspects may be surprisingly similar. Originally published in an academic journal, and based on historic fact, *The Anandrous Journey* is a narrative by the author from a compilation of family letters and The Birge—Merrill Correspondence, Part IV, page 125. At my family's request, the major portion of Harriet Bell Merrill's correspondence was placed in the State Historical Society of Wisconsin archives.

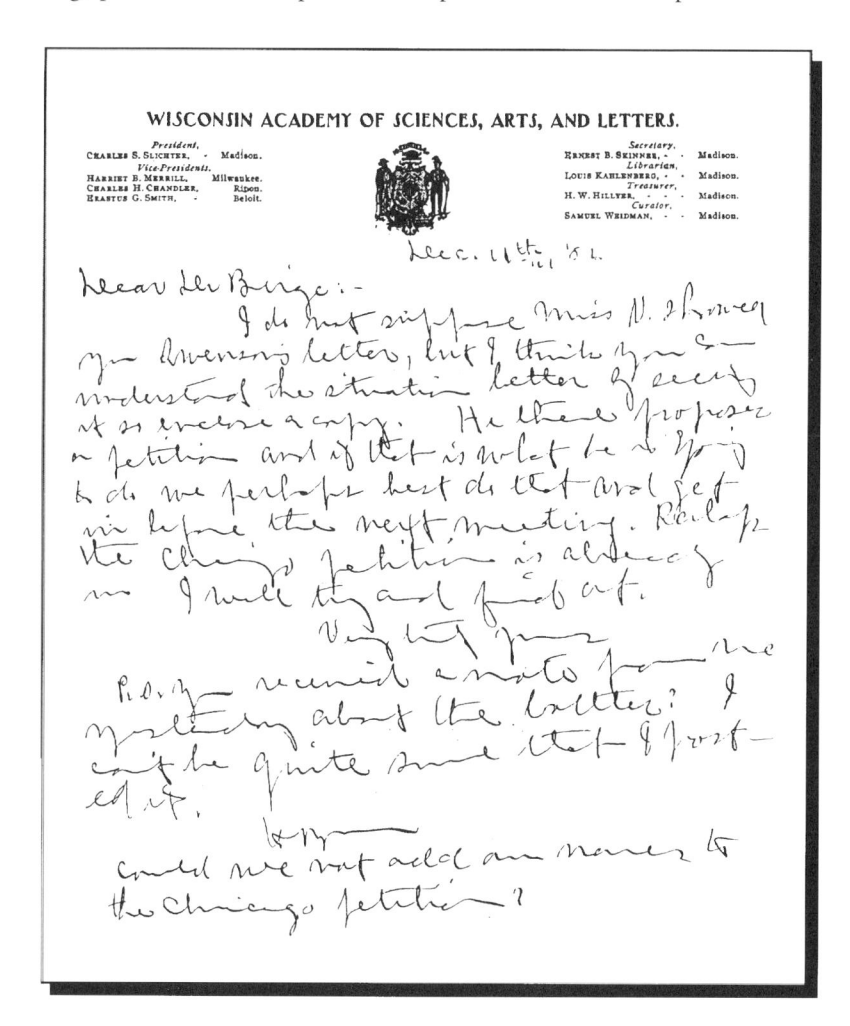

---

1. 1993 quotation of Emeritus Professor of Limnology Arthur Davis Hasler, as told to M. L. Hartridge.

© 1982 Lynn Hartridge

*The old Music Hall on Bascom Mall, original Engineering Building to the right. Merrill attended many functions at the Music Hall, formerly called Assembly Hall at the University of Wisconsin.*

# Part I

# A Biographical Narrative Based on Harriet Bell Merrill's Letters and Career as Research Assistant to Dean Edward Asahel Birge

*The Biology and Zoology Department, 1899. Photo courtesy of the State Historical Society of Wisconsin.*

*Woman at left considered to be Merrill wearing arm protectors during a laboratory experiment.*

April 25, 1901

Dear Cousin Nora,

*Nora was a member of Hattie Bell's mother's family and a teacher.*

    An answer to your inquiry about the Dean and my work is long overdue. You know how much I looked forward to accepting this position and the challenge of continuing research that goes with it. Remember when we were students? Our heroes were Pasteur, Semmelweis, Walter Reed and Marie Curie. She was one woman who persisted in her field of research. With all due respect and admiration, I wonder how she might have sustained quite the same intense interest in proving her theorems without the support of her physicist Pierre. I am told that Dean Birge has confidence in my ability to manage everything on my own and there are times when it is advantageous (when time is of essence). But Nora, so much of my time is solo! As for considering the Dean's personality, there is a dichotomy between the man and the scientist. He remains an enigma to me which seems to surprise no one.

    Edward Asahel Birge, the man, is small in physical stature, five-foot seven. I thought you had met him last fall. You can't miss his very full pate of graying hair, large drooping mustache and the discerning, deep set eyes glaring from under his bristling brows. If he were taller, I think I might find him absolutely formidable. Charles Adams (university president who proceeded C. P. Van Hise) claims that Birge is considered cold and unsympathetic but respected as a tenacious taskmaster. There are many, however, who openly admit that "His harshness approaches ruthlessness". Birge's statement, "All a scholar needs is one small room with a table, chair and a couch," certainly lacks warmth. He is known to show a temper in denunciation of opponents (during stiff faculty contests), which I have personally not witnessed. But if I have any hopes that praise from him might define the progress of my work, I am advised it would probably be too faint to be discernible![1]

    The Dean began work toward a master's degree in 1875 at Williams College, on the taxonomy of Cladocera. You may recall that when I was studying for my M.S. in 1891, he suggested it would advance his work if I were to pick up where he left off 13 years ago, with a comprehensive study on a particular species of Macrothricidae. As a result, I managed to isolate what was recognized as a not often found species of the genus *Bunops*.

    My work as an assistant professor is increasingly demanding with extra hours of lab for research as a result of "fishing" for Birge. He has developed a fine net, 1/20th of millimeter gauge, which is used to gather crustacea from varying depths in the lakes.

*Microscopic photo of* Bunops *less than one millimeter in length.*

---

    1. Refer to chapter notes, page 191: Reflections of Birge by Colleagues

ZOOLOGY                                                    249

tion is also given to the language reform movement. *Second semester; one credit.*
12. Scandinavian Literature. A general survey, with critical study of special periods. *Throughout the year; two credits.*

## ZOOLOGY

PROFESSOR BIRGE; ASSOCIATE PROFESSOR MARSHALL; ASSISTANT PROFESSORS HOLMES, WAGNER, MR. JUDAY, MR. SMITH, MR. GEE, MISS MERRILL, MR. WODSEDALEK.

### For Undergraduates and Graduates

1. General Zoology. The second semester of Biology 1, for which see index. *Five credits.*
2a. Invertebrate Zoology. A discussion of the structure, development, classification, instincts, and life histories of invertebrate animals. 2a is devoted to the lower invertebrates: the Protozoa, Coelenterata, Vermes, Mollusca, and some smaller groups. *First semester; lectures; M., W., 10; three laboratory periods; four credits.* Mr. HOLMES.
2b. A continuation of course 2a devoted to the higher invertebrates: the Crustacea, Arachnida, Myriapoda, Insects, Echinodermata, and Tunicata. *Second semester; lectures, M., W., 10; three laboratory periods; four credits.* Mr. HOLMES.
4. Vertebrate Zoology. A study of the structure, ph[ysiology,] habits, classification, and distribution of vertebrate[s.  ] *semester; lectures, Tu., Th., 8; laboratory and fie[ld work,] three periods a week; five credits.* Mr. WAGNER.
6. Variation and Heredity. A discussion of the main f[acts and] theories of variation and heredity, and their rel[ation to] other problems, biological and sociological. *Firs[t semes-]ter; two lectures a week; two credits.* Mr. WAG[NER.]
7. Evolution Problems. A critical discussion of the t[heory of] organic evolution, and the general development o[f evolu-]tionary speculation since Darwin. *Second semest[er; Tu.,] Th., 2:30; two credits.* Mr. HOLMES.

17

*University of Wisconsin catalog 1899-1901.*

*Edward A. Birge, University of Wisconsin, acting president 1901-1903, president 1918-1925.*

—*Photo from University of Wisconsin Zoological Museum archives.*

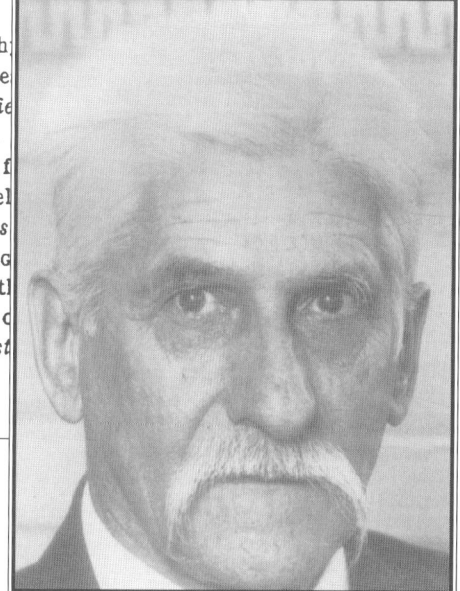

After capturing the animals from different levels, they have to be counted and separated according to location found. Out of 64 species found in Madison, 59 are from Lake Wingra. I've never spent so much time on lakes, and when I try to fathom why I pursue these tiny "water fleas," I justify that they are essential to the food chain of the world's water supply—therefore—LIFE!

Your Hattie Bell

P.S. When I am at Cornell I always hear high praises for President C. K. Adams.

∽

April 28, 1901

My Dear Hattie

You know that we thought you should remain at Downer as head of the science dept. As grueling as the schedule was, the seven science courses you taught were good preparation for your work at the university. The "girls" were eager for news of how you are faring. I was mum on the moot subjects.

I can well understand how your brothers and others in the family find your working hours confounding. Even the teachers here have a perfunctory concept of your regimen. Forgive me for scolding, but I think too much is expected of you. Can you ever say NO! Of

*—Photo from University of Wisconsin Zoological Museum archives.*

course, we are disappointed that you left Downer, but every last one of us is proud of your capability and spunk.

The curriculum is still the same. Pres. Ellen S.*[Sabin]* is as usual, holding reading sessions on Spencer, Payne, Kant and other philosophers whose work is currently popular to discuss today. Miss Wilder is teaching Elocution (which I was required to take) and Miss Williams is continuing in Harmony. There is no one who has been able to fill your "boots" in the hard sciences department!

If it is any comfort to you, there are many times when I wish there was less direction on what I can choose to lecture on in my Civics classes. I must be constantly on my mettle not to step on toes politically. The pleasure of your company is that your toes are not sensitive! Other than the courses mentioned, Downer requires its young ladies to take Penmanship and Etiquette, all necessary amenities considered "de rigueur" for a woman preparing to enter the 20th century.

The excitement on campus this weekend will be the May Day event, the annual May Pole dances around the old tree on Hawthorneden Green behind Holton Hall. I plan to attend. Come if you can. There is always a room for you. Remember, you are really never alone. You may not have Pierre, *[a science partner]* but consider this little clown I am sending as your "Pierrot". His smile should bolster you when you are low. Paintings of the character are the rage now and some of the students have cutouts of Pierrette and Pierrot and even costumes for parties.

Your Nora et Pierrot

—*From Milwaukee-Downer College catalog, 1900*

*Milwaukee-Downer girls celebrate May Day on Hawthorneden Green. Photo from old newspaper. Maypole tree and drawings of participants made by M. L. Hartridge.*

May, 1901

Dear Nora,

Dean Birge continues to promote the studies on zooplankton or aquatic protozoans and their effects on the thermocline of lakes and oceans. He is actively involved in furthering limnological studies (a new word in our lexicon) and has been using his position to solicit funds to requisition students of all disciplines to this science. The Wisconsin Geological and Natural History Survey is a contributor to the advancement of the Dean's research program, and he has hired on a biologist, Chancey Juday, from Indiana University who is expected to devote full time to the Survey.[2]

Although I am considered on an equal footing in my "boots" with male colleagues in my field, those of us assistant professors to Birge are of a sense that zoological studies are becoming secondary to the Dean's science priority—limnology. There is actually no definite directive for course proceedings, which can be disconcerting unless one is bold and innovative in their approach to research. This has left some students floating like so much academic flotsam and jetsam. Birge himself is taking few master's students and no Ph.D. candidates in the subject. He has no concept of my firm intent to continue this study toward a doctorate—if I have to go back to Ithaca or Urbana to get it! It is disappointing to think—with all of my related work—that this could not be accomplished here. Henry B. Ward, who is involved in the Great Lakes Biological Survey, has left Nebraska and will be in Urbana, head of the zoology department. He suggests I get my Ph.D. there.[3]

One of my colleagues in the department here related the story of how the Dean earned his Ph.D. Birge took advanced courses in Leipzig, studying theories of Gegenbaur and other German scientists. When he thought he knew enough about zoology to warrant the degree, he contacted an associate in Cambridge, Massachusetts, a Dr. Shaler. Shaler advised that the men on the board of examiners at Harvard wouldn't know anything about the German studies, and told Birge to make an application for an exam toward the degree, assuring him it would be an asset having a Harvard Ph.D. in a mid-western university. Shaler told him to "quote some zoological information, making a shot at Gegenbaur's work. If you don't know the answers the examiners ask, don't say you don't know," he advised. "Give them your version of your lake stuff![4]

Birge followed the advice of the influential member of the Harvard committee and was awarded the coveted degree in 1878. As you know, the degree would be most advantageous to me, but Birge hasn't pushed the thought.

I certainly don't have the confidence that Dean Birge had on acquiring a Ph.D. and I am aware of limitations that have confronted Professor Comstock who supported a graduate school. A couple of professors in Letters and Sciences are for it, but until the administration provides an adequate budget for operating expenses (with the consent of the

---

2. Refer to chapter notes, page 192: Chancey Juday—A Birge Associate
3. Refer to chapter notes, page 193: Academic Criteria—Early 1900s
4. Refer to chapter notes, page 194: Birge Earns Ph.D.

legislature), we will have to look for other options. After reading some of the regent's reports, I find it difficult to give positive answers of encouragement to my few female students who want to prepare for professions in the sciences. Particularly when the Dean still assumes that "men use their degrees as stepping stones to other professions while advance studies for women would be wasted." "Young women," said Birge, "look upon teaching as an occupation to fill the time between graduation and marriage." [1]

To some extent perhaps the men are correct. So few female students, even though capable and prepared, are willing to devote their entire lives to the sciences. Instead, for the most part they find themselves relegated to caring for families and the menfolk as well. Ivan Pavlov said, "The study of science can claim a man's whole life." To the extent that women in the sciences have to prove their intelligence and reliability through doughty independence, that effort alone can consume their entire lives!

This probably has not yet answered your question, "What is the head of the department really like?" Answer—next letter, I promise—or as much as I can fathom. Which reminds me of how I have spent the entire month, in addition to lectures. I have been studying the "water fleas" that were fished from the deepest levels. I am absolutely "bug-eyed" from pressing against the scope, hours on end while I make drawings of the remarkable crested chydorids! Not pretty, but fascinating. Meantime, let me know what is going on over at Downer.

Your bug-eyed

Hattie Bell


## Kappa Alpha Theta

Thursday, June 20, 1901

Dear Nora,

Another year—another commencement, and I've been here on campus to take it all in. Remember the time I came in from Milwaukee to attend the exercises and H. Siefert docked my pay when I was 15 minutes late in getting to my Monday morning class at East Division High? The train from Madison was delayed, with so many extra passengers returning from the weekend ceremonies, but I wouldn't have wanted to miss celebrating with my troops, that I had taught in high school, who had finally made it through the U.W.

I'm feeling a little like a student here at the Kappa Alpha Theta house. Ann Curtis, whom I knew at Cornell, submitted my name and I am a charter member. I was initiated in 1890 and must be the oldest member here. But age has its privileges.

---

1. Birge to Van Hise, Regents' Biennial Report 1907-08. For complete quote, refer to chapter notes, page 196: University Politics

I have a room to myself on second with plenty of shelves and a good lamp. "Befitting for an assistant professor to the Dean," according to the house supervisor. There is ample space to spread out test papers—my bed is made up for me—the toilet tray placed on a crisply starched dresser scarf is dusted and even the hair that I deposit in the hair dispenser dish is removed every day! I just hope it isn't returned to me made up into a switch, as a gift!

From my room I have a view of the front porch and fraternity row. I feel securely situated with accommodations far superior to what little housing is available for women here or at Chicago and even Cornell. You may have heard that the faculty, including Van Hise and regents, still think that Greek-letter fraternities are undemocratic and snobbish. After appointing a committee to examine the benefits of having privately funded housing, it surprised no one that fraternities were considered exclusive and that none other than the wealthy students were members. Indeed, they found that 27 percent of the men were partially self-supporting and 7 percent were entirely so. As for me, you know that I have been self-supporting and was granted scholarship status. Understandably I have not had time to participate in all of the events that take place in a social organization, what with meetings, lectures and research. When I am in town, I am fully occupied correcting papers in my pleasant quarters up on second—my "raison de entre" here at the Kappa Alpha Theta house. Professor Owen offered us this house at 630 Langdon Street.[5]

There is quite a commotion going on downstairs at present. Some Phi Delta Theta boys have strolled over and are serenading all the pretty young things planted on the veranda. Since the men are

*Harriet Bell Merrill.*
—*Photo from H. B. Merrill collection.*

---

*Sorority life—1890*

The Wisconsin Psi chapter evolved from a nucleus of six women students who had formed a secret society called the "Fault Correctors." Having selected their colors—their flowers—the grip and bylaws, they gathered every two weeks in room 21 at Ladies Hall. "Meeting in total darkness was required so that as criticisms were meted out, the blushes of those who were corrected would be hidden. Sometimes there were eats to help the girls onward."

Although Hattie Bell did not avail herself of the preliminary activities leading up to the Psi Charter celebration, May 30, 1890, she was present at the initiation held at the GAR Hall and at the banquet following, in the home of Father Leith. After breakfast the next morning the young ladies were taken on a steamboat trip around Lake Mendota.

from Catalogue of ΚΑΘ Archives—
*Sixty Years in ΚΑΘ, published 1902*

---

5. Refer to chapter notes, page 194: Student Housing and Greek Letter Societies

*A Lake Mendota steamer, Madison, Wisconsin.*
*—from M. L. Hartridge collection*

harmonizing on a sweetheart song, it must be a "pinning." This seems to be the same group that called for the girls this morning—took them on carriage rides around town and then a little steamer ride on Lake Mendota this afternoon. What is amusing is the leading lady in this role is all aflutter in a blushing pretense of absolute surprise over the whole ritual!

Well dear Nora, if you or I ever had such frivolous notions of engaging in events of that nature, we would soon face dismissal from our hard-won faculty positions. And should you decide one day to marry that dapper official on the Board of Education, he could stay on but you could no longer teach. Two wage earners on the same board are two too many! He is a fine gentleman, however,—so don't wait until you're too old.

Time to turn out this lamp. Have you noticed—when you look at the color green intensely then look away—you see red? I am seeing red. You asked me to describe the head of my department. I promise to do so as soon as I figure him out and when I'm not seeing red!

As Always,

Your Hattie

September 23, 1901

Dear Cousin Hattie,

    There is more news from the Board of School Director's meetings than from Downer in this letter. It seems there have been growing problems with delinquents in some of the Milwaukee wards. If it weren't for the Women's School Alliance and the Aid Alliance, most teachers would have their hands tied. These leaders have urged the board to authorize teachers to use strict disciplinary measures with the most disruptive pupils and have ruled that truancy will not be tolerated. In some districts children no more than 10 or 12 years of age are made to work during part of the school day. They are usually the ones who are the truants and I am sorry for them. We have also been authorized to remove from the rooms those students who repeatedly show behavior that disrupts the entire class. I have physically had to remove only one boy from a class, sending him to the principal who is known to use a lilac switch. His method is reputed to be effective in the extreme cases that threaten authority. Remedial institutions have been considered the only solution for the unmanageable delinquents. They not only waste the teacher's time but slow the learning process of students who are bright.

    Occasionally I monitor civics classes at S. High, and I've noticed lately almost as many Polish, Russian and Bohemian youngsters as Germans in the school. The Alliance rule, in accordance with the Women's Relief Corps (Drake Post), makes certain that the American flag is saluted and the Pledge of Allegiance is recited every morning. We are all proud of the history department. The board is justified in requiring that the history of our country and its leaders be taught to the children of immigrants. They should understand our laws and the leaders who made this the country that they have chosen to come to. It did my heart good at the Board of School Director's meeting to hear the comment that "the kind of freedom enjoyed by Americans is impossible without the discipline that respects the life and property of others."

    Thadeus Wild wants Polish history and the Polish language instituted in the grades through high school. Edward Rissman agreed. Most of us responded that other language courses should be available only at the university level. Eventually every other group will want their separate, individual studies, and we will end up a fragmented society. True, we have more German-speaking teachers than other languages, although Bernard Abrams emphasized that the Milwaukee State Normal School be requested to strengthen the

*This photo is thought to be of Nora.*

—*from H. B. Merrill collection*

German department, making certain there are sufficient teachers prepared to teach German right on down to the primary grades. I certainly have not noticed a shortage of teachers who are equally conversant in German and English.

Most of the German-speaking teachers I know in Milwaukee are promoting English as the major language and are fully aware that it is essential in the world of business. With too many languages, there are conflicting interpretations. One language makes for a united citizenry working together—it seems logical.

It has been noted that the German communities are solidly for hygiene and gymnastics, which have been promoted by the Turners. In some wards, the schools have no showers for the girls, and children from certain districts have no bathing facilities at home. Some German leaders concocted showers in the S. High School basement. They collected barrels from a cooperage and rigged up water pipes from the city main to fill them. Suspended from the rafters, holes in the barrels release the showers when tipped. In many cases, these are the only baths the children get—once a week if they are in gym class. We can also thank Mrs. Silas Merrill (your shirttail) for her influence in promoting better lighting and ventilation in the classrooms.

Do you recall when you were at E. High, there was one towel per room for two weeks and one drinking cup hanging by the spigot? Now there is a theory about bar soap carrying germs. Good old Ivory has been replaced by a tilting "vase" that dispenses a milky liquid (for $2 each!). There are also more towels and cups! Remember old Gus Hering? When the Merrill group noted that the dry-sweeping of schoolroom floors was contributing to nose and throat irritations and perhaps the incidence of lung disease—even consumption—it was suggested that the floors should be wet-mopped at least once a month. Gus reasoned to the board that the students were asked to "clap" the chalk dust outside of the classrooms, so there wasn't much to sweep up indoors. Though some of the janitors are receiving as much pay as teachers, and many teachers are paying extra out-of-pocket to have janitors clean their rooms, Gus said he would wet-mop only for extra pay. Well, the board said the budget couldn't sustain the cost—so, no wet-mopping!

The subject of hygiene brings to mind bacteriology, a discipline you are familiar with. The Women's School Alliance was responsible for cleaning up unfit water supplies, particularly in District 9. We have attorney Austin to thank for forcing the board to consider inoculations of vaccines throughout the public schools. It is the most effective method of controlling communicable diseases among children and families. As usual, there was objection among the uninformed. I will never forget 1897 and '98 and the outbreak of diphtheria. The schools were closed for months. San Hooper was paid his total salary for the year, $2,400. Our amount, which was exactly half of that, was considered leave of absence without pay. One of the worst years was 1898. First the loss of 260 of our poor Navy men on the *Maine* and then the war. Every flag in town and in our classrooms was at half-mast for the duration.

Incidentally, my seniors are studying the presidency, and I put the photograph of McKinley (you loaned me) up on the bulletin board. I explained that you were in Buffalo, New York, at the Pan-American Exposition and heard his talk on September 6, 1901, the day he was shot! The students were quite impressed with both the photo and the poster of

*Buffalo, New York, Pan-American Exposition, September 6, 1901. President McKinley on podium. H. B. Merrill took the photograph standing just behind the person with umbrella at right.*

*From Merrill's collection of personal items. Original poster copyrighted in 1899 by Pan-American Exposition, Co.*

—*from M. L. Hartridge collection*

the Pan-American Exhibit. With all of the advanced x-ray equipment that was displayed on the grounds, it is still puzzling that medical attendants didn't use it to locate just where the bullet had lodged in President McKinley's back. He might still be alive if they had found it!

Now dear Hattie, you are about to walk into danger as well, albeit a different kind. The countries you are going to are rife with the most virulent diseases, many of which I suppose are endemic to natives of the various regions. What is the concept in South America regarding inoculating the populace and how in Sam Hill will you be protected against the bubonic plague, cholera or yellow fever?? You never gave St. Christopher any credence, but we will pray—God be with you.

Nora

P.S. There is a request here for a live armadillo, which it is hoped you might bring back from the "jungles." Most of us see them only when they are dried, varnished and made into Christmas gift baskets tied with red ribbons!

∞

November 18, 1901

Dear Nora,

You are a reliable confidant although what I relate is common knowledge throughout town and gown. You are merely an earlier recipient of the information. You are aware of my additional duties, and I hope will acknowledge that it is more propitious for me to write less often but more at a time. How is that for justification of my "missives"?

Now that he is acting president, Birge is spending less time on research and does no teaching. I am totally emersed in aquatic studies in the laboratory, lecture sections and at night paper work. Chancey Juday, whom the dean hired as his biologist for the survey, is finding it hard to keep up and has not been well. I work at a pace that accomplishes what is required, but it is becoming tediously routine.[6]

We are aware that the Dean rarely misses his Literary Club, Whist or church services, although we have learned, from those who can attest, that his wife Anna has to nudge him awake frequently nowadays. He turns over his salary to her and has insisted that she have a full-time maid to help her entertain as her position now demands. They have had parties and receptions with orchestras and refreshments up in Library Hall. Some of the students here in the department were joking the other day about the Dean's latest "field project." It seems he had to repair a tennis court that he marked out on the back lawn! If it were for young Ted and Anna, I am touched.[7]

---

6. Refer to chapter notes, page 196: University Politics
7. Refer to chapter notes, page 198: Birge's Private Life

Birge is very short on small talk, so I learn about the family secondhand. However, he does use me as a "sounding line," *[limnology term]* as he calls it, when he wants a reaction to what the faculty might be thinking on current political issues as they apply to the UW or the Women's Self-Government Association. He would feel too vulnerable to reveal the need to know from a male colleague, I'm sure. The failure of President Adams to recover from a breakdown caused Adams to resign in October. Senator Vilas is chair of a regent committee commissioned to find a new president. Many men from other universities have been considered. It seems only logical that the acting president, who is already entrenched in the position, should have serious consideration. Adams himself seems to be agreeable to Birge. I have said nothing about dissidents among the faculty, but he knows.

Professors Turner, Slaughter and Slichter, who are friends of professor Van Hise, are urging his election. They are politically very influential, even being called the "Francis Street Cabal" and "the King Makers." Also Governor Robert LaFollette, a fellow classmate of Wisconsin, 1879, feels that Van Hise is the man for the office. Although we as women are not supposed to concern ourselves with voting on issues of any importance, I will launch a petition among all alumni groups, advocating Birge for president. As Vice President of the Wisconsin Academy of Sciences, Arts and Letters and of the Wisconsin Alumni Association and every other organization I represent, I am drafting a letter. Henry Vilas will support it. *[See page 105: Harriet Merrill's Return to Appointed Tasks]*

You are probably wondering how I can support a man who has stated in papers (and I feel still believes) that women can not be expected to make scientific research a lifelong career. *[Refer to chapter notes 6, page 196: University Politics]* The only encouragement I have received from him is that he said that I was intelligent enough to proceed on my own! Am I just grateful that a learned man has acknowledged that I have a brain? Didn't Milton say, "To deny being vain is a coy excuse?" Something like that. To answer my own question regarding Birge—I think he and his family will be deeply hurt not to move into the position he has prepared for.

I must now get a letter off to the Milwaukee alumni and my connections at the University of Chicago who could use their influence with the electorate.........so to bed,

Your Hattie Bell

∞

Dec. 11, 1901

Dear Dr. Birge:

I do not suppose Miss N. *[Northrup]* showed you Swenson's letter, but I think you can understand the situation better by seeing it. Enclosed is a copy. He therefore proposes a petition and if that is what he is going to do we perhaps best do the same, and get it in before the next meeting. Perhaps the Chicago petition is already in. I will find out.

H. B. M.

P.S. Could we not add our names to the Chicago petition?

January, 1902

My dear brother Roger,

If you wonder why you haven't heard from me lately, it's because the work load in the department has increased since Birge was appointed acting president. Charles Adams has taken a leave from the university due to ill health. He was well liked at Cornell by the way. He will go south and then possibly to Europe.

Meantime, I am actively involved at the Wisconsin Academy in sciences and have, within this past year, succeeded Frederick Jackson Turner as Vice President of Letters. I have been impressed with papers he has presented on "American History As It Has Been Determined By the Natural Environment." Incidentally, I have been proposed as a member of the Wisconsin chapter of the Natural History Society and told that election is assured.

You would find the debates here on conflicts between religion and science stimulating. For Birge, to arrive at convictions concerning evolution seems to present a dilemma. He is a disciple of Saint Paul and has arbitrarily initiated a series of studies of Paul's letters, which he discusses at the First Congregational Church. He has described science as "A construction, which developed as a place where human thought might work, where the spirit might live and a temple where it might worship." He continues to elaborate upon his religious theories with an almost Darwinian philosophy at times. *[Refer to chapter notes, page 198: Birge's Private Life]* Was it Voltaire who said, "If God did not exist, it would be necessary to invent him?"

When I focus on the most minute microorganism through a 1/250 power microscope and then view galaxies of "celestial cells" through a telescope, I feel that all material in the universe originated from the same basic cellular synthesis which varies only in scope and purpose. All of it is a part of the Almighty, and we on this planet are but a microcosm of the basic, infinite structure of the whole. I have maintained this hypothesis since our "tadpole days," and I am content with it though you may think me daft.

I often think of our excursions through the woods along the Wisconsin River. I was the one who collected insects, rocks, and plants, taking them home for closer scrutiny under magnification. Then I would leave them scattered among Nate's neatly arranged toy soldiers and all of your instruments, the purposes of which I considered less important than my unique specimens, hah!

We weren't directly affected by the War Between the States, but those were not hal-

*Harriet's brothers—from left: Roger and Nathan Merrill.*

cyon years for Mother and Father. Whenever I remember that ponderous Victorian structure we called home, in Stevens Point, I recall Father's remark, "The war built it," he said. It was all the timberland opening in Wisconsin that brought him from his mill in Brunswick, Maine. I remember he said that he "had to build enough room to house all the books his schoolmarm wife refused to part with!" Losing Father in 1880, and now Mother, has set me to reminiscing.

Come see me when you can. I am not able to leave for the foreseeable future.

Love,

Hattie Bell

*—from H. B. Merrill's book on studies of* Filicinae

February 27, 1902

Dear Nora,

I have been getting advice, from sources not even queried, about my pending trip to South America. You my dear Nora, who in some measure directed or rather restrained my longing for more extensive travel, have said, "Go!" I have always found your advice so reliable. The dean, with his quick perception of what I might collect in crustaceans for the department, assured me that he would see that I had an adequate supply of his finest silk nets. Can't you just picture that…me with my camera equipment, boxes with slides, and bottles for specimens, slung across my five foot tall, one-hundred-pound frame? Our mutual friend, superintendent of schools over there in Milwaukee, says I am a brave woman and he would like to be going along. He used to teach geography you know. The head of the Board of Trustees of the Milwaukee Public Museum is extremely concerned for my safety in such a vast and "hostile" environment, although they all grow more excited each day as they contemplate what acquisitions might end up in the museum as a result.

UW Professor John Freeman, who cruised the Caribbean earlier this year (he's the one who collects butterflies), has given me the most practical bon voyage gift—Persian insect powder! He instructs, "When the pests get too troublesome, make a high barricade around the perimeter of your bunk and get in the middle of it." As for my brothers, they have told me in a manner of reprimand that my arrangements to go are irresponsible. When I asked, "To whom?" Nate said, "It is entirely out of the question for a petite little woman to hazard such a rigorous venture virtually on her own…we will not discuss the matter further!" Roger pleaded with me to wait until next year and he would go with me, if for no other reason than to carry a gun. I never did learn to shoot properly.

In spite of all the deterrents, I am ready to roll to Rio. I read Mary Kingsley's article on West Africa, but find few recent accounts on South America, where she did not venture. Instead, like Kingsley, I am filling up the vast cavities of my mind with geographical and historical statistics on the subject, which I am instructed to codify for the general readership of the *Milwaukee Sentinel*. I will post my personal unedited perceptions on to you as usual and you can follow the rest in the paper. If I don't get to Milwaukee before heading out to New York, give my love to the girls at Downer and E. High and keep some for yourself.

    Your Hattie Bell

P.S. I was about to go out and post this note when another well meaning housemate gave me still one more strong warning about traveling alone. She reminded me that I am only six months younger than Miss Kingsley and the jungles are not a place for women in their 40s. Kingsley did not get to South America, but did go to Africa where she joined some of her countrymen in the Boer War to help the injured. She took fatally ill with a tropical disease and before dying, asked to be buried at sea rather than in her homeland, England.

I am intent in going solo on this journey as I have on all others, *but I will come home*, one way or another!

June 10, 1902

Dear Roger,

We just had word that Chancey Juday, Birge's key assistant who was diagnosed to have tuberculosis, will have to be away another year or more. He was here only one year after Birge hired him from Indiana. He has been at the universities of Colorado and California where he hopes to regain his health. This has been a great loss to Birge as Juday is exceptionally knowledgeable. I never got to know him very well. He worked solely on diel migration of zooplankton in the lakes of southern Wisconsin. *[Refer to chapter notes, page 192: Chancey Juday—A Birge Associate.]* I had just lost Mother and had no time off from teaching. I still haven't taken care of all of her personal things. Nate took care of other business matters. So far, Birge has not selected anyone to replace Juday here, and I will not neglect my students.

The Dean himself should have the time now that his presidency was relinquished. *[C. P. Van Hise, well respected and noted University of Wisconsin geologist with the backing of Governor Robert M. LaFollette and professors Turner, Slaughter and Schlicter was unanimously elected president of the University of Wisconsin in 1903.]* Dean Birge is planning to send his Anna and the children off to their folks out East for the summer as he does every year. He claims it affords him more time for lake studies. I was to do summer school classes but am definitely going to take off and head for a warm climate and change of pace. It has been an exhausting winter. I have contacts at Cornell and Chicago who, when they learned that I was interested in going to South America, assured me that I would recover expenses if I were to lecture upon my return. So, I announced to the Dean that I have made arrangements to research on my own—out of the country! Details anon,

Your Hattie Bell

P.S. I have been told that there will undoubtedly be a hiatus in the limnology department with Juday gone a few years, and I plan on making two trips to South America, so we will see how progress will be met on research!

*Captain Leonard Parker Merrill, youngest son of Roger and Sarah (Freeland) Merrill. Born in Brunswick, Maine, September 29, 1821, and died in New Orleans, Louisiana, October 31, 1890.*

June 24, 1902

Dear Roger,

I was so glad to get your card from New York and know that Mr. Warfield gave you some leave from the home office in Baltimore to attend the Rudyard Kipling book review conclave. We have heard here that Mr. Kipling had a falling-out with certain family members, and saddest of all, that he and his daughter got pneumonia while there. I read that his daughter died.

You know how much I enjoy his writing and this latest, "The Just So Stories", have me marking time with an invigorated cadence. I appreciate receiving the book, but after reading the lines in "Beginning of the Armadillos," I can't get the poem out of my brain.

I'd love to roll to Rio someday before I'm old! I'm not going to blame you or Mr. Kipling for getting me in this frame of mind. I have dreamed for years about going to South America, and now my mind is made up and I must roll down to Rio if I am to remain able to put up with the roll in order to enjoy the wonders. I will let you know soon about my arrangements.

I am looking at the picture of our uncle Captain Leonard Parker Merrill that Aunt Bella *[Arabella Merrill]* gave me. If he could ship/mast from Bath *[Maine]* and navigate one of his ships around Cape Horn, I think I can take the rigors of reaching Tierra Del Fuego as a passenger.

Your "old" Hattie Bell

New York

July 2, 1902

Dear Dr. Birge:

    Your second letter came just as I was about to leave. I am so sorry to trouble you. I have not read your notes. Will read on shipboard.

    Schoff was very nice to me, giving me more than a dozen personal letters to Pará, Manaus, etc. I shall certainly go to Manaus. I have a whole stateroom to myself—only about 20 passengers. Congressman Burleigh of Maine is one of them.

    Well, good-bye for now,

Yours very truly,

H. B. M.

*Just So Stories*
by Rudyard Kipling

## The Beginning of the Armadillos (1899)

I've never sailed the Amazon,
    I've never reached Brazil;
But the Don and Magdalena,
    They can go there when they will!

Yes, weekly from Southhampton
    Great steamers, white and gold
Go rolling down to Rio
    (Roll down—roll down to Rio)
And I'd like to roll to Rio
    Someday before I'm old!

I've never seen a Jaguar,
    Nor yet an Armadill
O dilloing in his armour,
    And I s'pose I never will,

Unless I go to Rio
    These wonders to behold—
Roll down—roll down to Rio—
    Roll really down to Rio!
Oh, I'd love to roll to Rio
    Someday before I'm old!

The cache of letters written by Harriet Bell Merrill to her friends and mentor at the turn of the century, were fragile, like the botanical specimen that long ago had been pressed between pages of a favorite poem by Rudyard Kipling.

*Map of South America from Scholar's leather-bound Edition of 1910 Encyclopedia Britannica Company, Eleventh Edition owned by D. John Leigh.*

# Part II

# Journey Through South America

*Photo is from a tributary of the Amazon on a trek to the interior, July 23, 1902.*

# Journey Through South America by Enterprising Milwaukee Woman

*Article to the* Milwaukee Sentinel *published February 15, 1903*

February 15, 1903

As to going to South America on my own, I broached the subject tentatively to friends and associates. The superintendent of schools (the one whose pet hobby is geography) said, "You are a brave woman Harriet." Dean Birge, with his quick perception of the end rather than means, said, "Excellent idea—good chance to collect crustaceans." The director of the board of trustees at the Milwaukee Public Museum scorned the idea of a woman five feet tall, weighing one hundred pounds, to collect anything from the vast wilderness of the southern continent.

My brothers told me it was entirely out of the question to go without one of them accompanying me. One of my young assistants warned that the climate is fatal to white people and added that Wisconsin folks were the whitest he had ever seen. The time spent in convincing friends that I intend to go left me little time for preparation. I have packed a sensible suit, a good dress, my camera, nets and notebooks. My boater, a knockabout hat and my boots complete the wardrobe. John Charles Freeman, who professes linguistics and collects rare butterflies, cruised the Caribbean Sea earlier this year gave me important advice and a valuable bon voyage "gift." He said to take plenty of Persian insect powder and when the pests got troublesome, I should make a high barricade of the powder all around the perimeter of my berth and then get into the middle of it!

> "Leaving the middle western university scene and sailing to South America was a release as liberating as loosing my corset stays and changing to a shift." —H. B. Merrill

*Article to the* Milwaukee Sentinel

# Merrill Really Rolls to Rio
# Sailing Out of New York on *S. S. Byron*

The *Byron,* on which I sailed, was supposed to be the best of the line but was dirtier and more uncomfortable than any ship I have ever sailed on. Huge cockroaches (with a long roll to the "r") were swarming over the ship. I made a high barricade of Persian powder in a circle and got into the middle of it. If John Freeman could succeed with his six-foot two-inch length, I was sure I could ward them off. The powder shifted with the roll of the ship as I sat there holding my nose.

Bilge water and cockroaches are odious enough but when combined with insect powder, the situation is unbearable. The roaches merely moved to a different and, if possible, worse place. They crawled on my berth and over my head. They got under my pillow and when I turned it over to the cool side, they scurried away like creatures with guilty consciences, antennae waving like pitchforks waiting to return and impale their victim in the night. They formed a mosaic in my washbowl. They crawled into my teacup before I could lift it to take a second swallow. The South American passengers, with native pride, assured me they are nothing compared to the giant South American varieties, as large as mice! They might be more agreeable than the millions of croton bugs on the *Byron,* which have no respect for repellents of any kind.

I prefer to spend more time up on deck away from the pests and the odor of bilge water and find it invigorating to watch the flying fish, whales and Portuguese man-of-war, in spite of the gales, knowing that rest and warmth are assured before long. There has been some entertainment as we crossed the equator. The Captain permitted considerable license allowing the sailors to play tricks on passengers who are crossing for the first time. I almost felt compassion for the unwary newcomers who were lathered and shaved with barrel staves, rolled in flour and then dunked. They looked like giant cutlets ready for deep frying. This is the third week that we have been out of the sight of land or any other craft, but the weather is mild.

Titanus *beetle (actual size), South America.*

South of the equator, the constellation Southern Cross looked neither bright nor large to me, and the stars are arranged more like a kite than a cross. The Brazilians use it as a national emblem on their flag and money. When passengers on board asked my origins I had to answer in my limited Spanish, "Estados Unidos de Norte America." Their reference to me as Professor is reassuring as they seem to understand that my purpose is scientific research. Most of the passengers are Spanish and Portuguese speaking men, Canadians and a couple of Americans. There are only three women on board and I am the unmarried one, and the one wearing what they call a hunting outfit and carrying "fishing tackle!" My man's shoes, which are better constructed for combating unpredictable terrain, are reliable on an unsteady deck. We are really rolling and I am surprised that I have not been seasick.

*Report to the* Milwaukee Sentinel *and The Milwaukee Public Museum*

# Arrival in South America

### S. S. BYRON

My worst fears of going to South America are confirmed, when our first stop is Pernambuco which is rife with bubonic plague. I feel more foolhardy than brave at this point. Quarantine flags fly at most every port. Bahia is the first city where we are permitted to enter. It is a typical tropical city and one gets interesting first impressions from its people and its location with steep narrow streets winding up hill to a crest on top.

From the shore, the houses look dazzling white and it seems a spotless town, mayor and all. The men are fond of uniforms if they are officials of any status. The illusion soon vanishes however when one gets into the business part of town.

Passengers go ashore in rowboats because of lack of wharves. Freight is handled in lighters. The streets are full of idle, lounging people looking like large dark bronze statues who crowd about offering all sorts of unneeded services. The houses are made of brick covered with plaster tinted pink, blue, green, red and purple with red tiled roofs. The whole effect is very cheerful. The absence of chimneys is obvious to a Northerner. All of this color against a deep brilliant hot blue sky is glaring in the sun but attractive in total effect along with the almost costume look of the native's attire.

The great open markets, overflowing with fruits, vegetables, fish, flowers, clothing and jewelry for sale, are like great fairs. The oranges and sweet lemons, guavas and pineapples and even strawberries are available at the same time. Parrots and giant butterflies loll about as lazily as the people. Except for the dock area it is difficult to see what endeavors the populace is engaged in.

*Bahia as seen from the water.*
—H. B. Merrill photo

*Unloading cargo.*
—H. B. Merrill photo

*Natives in marketplace at Bahia.*
—H. B. Merrill photo

*Woman from Bahia carrying laundry on her head.*

—H. B. Merrill photo

    I hope I am getting some good photos of these great dark people. *[See photos on pages 95, 96.]* It is the women who carry everything on their heads, who are the graceful ones. They prefer to wash in their own chosen depressions in rocky places along the river.

    Laundry work is done in the most primitive fashion. Clothing is taken to brooks, rivers and ponds and washed. It is then boiled in the open air in tin kerosene cans. The negro women bending over these cauldrons look like so many "witches" in Macbeth. When washed, the garments are spread on the grass to dry. Washing is not a weekly affair, but a continuous process. I do not know how they divide the ponds, but groups of women are always washing, and the water is always cloudy with soap suds. On the whole the poorer classes even give evidence of this continuous wash day. There is a far greater general average of external cleanliness than the northerners expect. The women wear their clothing starched so stiffly you can hear the rattle of their skirts a block away.

    One can not but be impressed with the wastefulness of the above method, both in soap, fuel, and labor, and the destruction to clothing, which comes back faded and torn, and I often wondered if our steam laundries could do a paying business if once opened. I suppose not, as fuel is expensive. Only the wealthy could afford to patronize them, and every family has a laundress that seems to prefer the old method. These tall statuesque women seem to stake claim to certain bowl like crevices eroded into the rocks probably from generations of the practice. From the laughter and singing that accompanies the task, it almost appears to be a competition for the best performance. When the laundry is dried, it is piled high on top of their heads without a spill. One almost feels like applauding!

    H. B. M.

*Merrill's personal letters have similar content but the information varied to different persons and in articles to the* Milwaukee Sentinel

S. S. Byron
July 21, 1902

Dear Dr. Birge,

    My worst fears of South America were confirmed when the first attempt of my ship to dock in Pernambuco was forbidden because of the bubonic plague, which was everywhere in the city. Quarantine flags and the Yellow Jacks, cautioning entry, were hoisted at several ports. Newspaper headlines shouted, "All Brazil—is shunned as the land of death!" During the period when I was there, hundreds of people had died of bubonic plague and several North Americans were hospitalized.

    I am beginning to think I'm a fool rather than a brave women. At Bahia, the ship's doctor (McGuire) took me ashore and way out to the interior where I gathered samples of sediment from inland lakes. Wherever I land the threat of cholera, malaria and the plague are prevalent. The intermittent riverboats manage to deliver mail, though unpredictably, but I will continue to correspond as time and connections allow.

    Yours truly,

    H. B. M.

*Dock at Pernambuco with boat from Bahia shown.*
    —*H. B. Merrill photo*

*Article to the* Milwaukee Sentinel

# Flora and Fauna
# and the Botanical Gardens of Rio de Janeiro

We approached Rio with some trepidation after learning the plague is prevalent, but got up early because the captain claimed that, aside from the harbor at Sidney, Australia, it is the most beautiful harbor in the world. Going through a chain of islands the boat entered between "Ma" and "Pa" to the large islands, bordered by rows of mountains. The channel is so deep and the boat so close to land that we could almost touch the trees as we glided past. In some ways it reminds me of Portland, Maine.

Butterflies in clusters looked like giant blossoms fluttering—a paradise for my lepidopterist colleagues. Rio is everything I expected. Huge tropical birds with brilliant plumage greeted our boat up close.

They seemed to know that even if I had a net, I would not capture them for a display case. The flowers, too, appeared to have escaped the conservatory, allowing them to expand to enormous size and vivid coloring. Lewis Carroll would be inspired! My only disappointment was the Avenue of Royal Palms. Considered the finest example of the species in the world, they look like enormous feather dusters stuck up out of the ground. The ones at Barbados are more graceful, but those in Rio are too high in proportion to their diameter. Seen singly, with the lower dead leaves whipping about in the wind, they look awkward. No variety of statuesque tree can compare with the North American giant redwood.

There is an interesting story behind the creation of the Jardin Botanico in Rio. It seems that in about 1807, Napoleon sent an army to conquer Portugal, which was still allied with Britain. When it appeared that Portugal would be taken, the royal family sent mad Queen Maria I and her son, the Prince Regent, to their colony in Brazil, escorted there by a British naval ship. They immediately set about to beautify the grounds of a 16th century house and a sugar mill that was surrounded with luxurious tropical vegetation. It was the Prince Regent Pedro himself who planted the palms that have been called Royal since that time. The roots of

*Royal palms at The Botanical Garden, Rio de Janeiro.*
—*H. B. Merrill photo*

the palms were brought to the Regent by a shipwrecked Portuguese naval officer who had been a prisoner of war after his capture by the French. Thrown in with Africans, Indians and Chinese who were sold as slaves by tribes off the coast of Africa, Luiz Abreu (from Goa on the west coast of India) became fascinated with the rare plants in the area where he was held. He gathered roots and seeds from trees and flowers before escaping in 1809 to Brazil, where he presented them to the Prince Regent.

I must admit, I have not seen more exotic specimens than these descendants of "royal contraband," but there are many who are responsible for the survival of this nationally known botanical garden, other than Brazilian nobility. There was a Carmelite monk who was a botanist as a Fellow of the Horticultural Society of London. He adopted the Linnaean system of plant classifications and was influential in establishing a botanical center, which was financed by the son of the Emperor Dom Pedro II. Those who took me around the gardens gave me a history lesson along with the horticultural background. They seemed proud of Pedro II, whose memory they revere since his death a dozen years ago. But if you talk to the large landholders, they will tell you that he was too liberal. He attempted to abolish slavery and that prompted the coup that made Brazil a republic.

Dear Nora,

As you can see by the post, I am in Buenos Aires.

On one side of the Río de la Plata more than 60,000 inhabitants exist in the miasma of La Boca near open drainage systems, worse than France. Fecal material from highly congested conventillos runs down the street, and houses of questionable activity are next to orphanages which are also in proximity of churches! In contrast to this poverty on the west bank, is enormous wealth on the other side in the city of "good air." In the grand homes, servants stand at the front entrances to show guests into the "sala," or reception room. The English are very correct in social protocol here but have long since stopped putting their coachmen in livery, particularly the ridiculous looking "Billycock hat." Some of the British customs that are now passé are carried to extreme by South Americans, much to the amusement of North Americans. One thing that cannot escape notice is the warmth of the people. It equals the climate. No matter how many times a day you meet, men and women shake hands, bow and often kiss one another on both checks. To me this is a trying ordeal.

Your Hattie Bell

*Policeman in Buenos Aires.*
—*H. B. Merrill photo*

# Garden of Decadence

Dear Cousin Nora,

    I had looked forward to the botanical gardens in Río which were truly exhilarating. On another trip, the gardens in Montevideo, seemed minor in comparison. Perhaps there should be no comparison between one or the other any more than one can compare a wood violet with an orchid. I had such an excellent guide at Río. In Montevideo, a visiting professor at El Prado gave me a perfunctory tour of the grounds, emphasizing the statuary rather than the botanical specimens. I wondered if he assumed that female science professors thought only in terms of inorganic material as he walked me past fountains filled with lilies over to a bower peopled with stone figures. As if he were a museum curator, he proceeded to point out, in a most repugnant manner, all of the physical attributes of the female nymphs and muses. I told him I had seen the figures representing Art, Music, Poetry and Science in Río and if there were a statue of a woman scientist, I was interested.

    Then the overbearing Englishman (for whom foreign translation wasn't needed) said that he was informed that I was on a research sabbatical and was curious as to "why a genteel lady of my delicate nature was traveling alone?" Ignoring his ill-bred questions, I calmly explained that what I was hunting was an animal too small to be seen. Thinking this would puzzle him and settle his inquisitiveness, I turned to go, when he apologized for not having a monument to a scientist in the garden. I thought he was guiding me out when he led me toward a rather dark arbor where a fairly large statue of woman with a broken bow and arrows stood. Forlorn in such a derelict condition, she looked like a fragment from an abandoned culture.

*Artemis "Diana of Versailles," a statue of the goddess "of the hunt."*

    "This is Diana, the huntress," he said with a wide sweep of his arm in introduction. "She, like you, is a goddess of the forest who hunts for that which cannot be seen!" If he hadn't said it with a laugh, I would have had to stifle my own incredulous snort. The thing that did not amuse, but repulsed me, was the way he began to remove the moss from the statue, almost sermonizing about how the beauty of the human body should not be hidden and that removing the mold was symbolic of taking away shame!

    Thank heaven a couple of caretakers walked into the area where we were. If an anaconda had dropped down from a tree in the garden, I would have been less apprehensive than I was with that perfidious viper! Would you have reported his behavior? Since he didn't touch me, I didn't. I wish that there had been a Carmelite monk to walk me about, or at least a statue of Nemesis, the goddess of vengeance!

    Your Hattie

    P.S. I got some good photos of the Royal Palm Avenue in Río.

São Paulo
July 31, 1902

Dear Dr. Birge:

    I have not written you since my arrival on terra firma because my letters would not go out until the *Byron* returns to pick up mail. The Purchases met me at Rio Thursday, July 22nd on Colonel Byran's government launch. My trunk did not go to the Custom House at all.

    I had a delightful time with the family at Petropolis doing the "social act." There was a reception at Colonel Bryan's, breakfast at Consul Seeger's, etc. Went fishing but found nothing so came back to São Paulo as Mr. Purchase is opposed to my staying even one night in Rio. The yellow fever is fatal and the English and Americans steep themselves in whiskey. Yellow Jacks fly at almost every port. I go from here to Santos and to Buenos Aires on the fifth—back to Rio and then to Pará. Your mail to me should go to Pará in care of the London and Brazilian bank, where my other mail is received.

Everyone has been very kind to me. Consul Seeger says I should be able to go to Pará. He arranged for the delegation for the St. Louis Exposition to bring me on here. A wealthy American, Theodore G. Sullivan (Cornell graduate), is traveling with his wife and daughter (a sophomore at Wellesley) to Buenos Aires on the same boat I am taking. They have invited me to stay the night there before sailing. People seem so glad to see Americans, they will not take my pay for accommodations. Yet everything here is very high. Mangoes are $1.50. Oranges elsewhere—a penny a dozen.

As to "fishing" and returning the specimens to you, I have done quite well. I am not getting the species I need but have instructed my Guarani boy to help me carry equipment if needed. I saw Professor Derby, who knows Dr. Steven M. Babcock very well, and Mrs. Wells, a Cornell graduate, knows him, too. I have collected mud samples, which are being sent on by Dr. McGuire, the *Byron's* ship doctor. I cannot send heavy bags by mail of course. The doctor, who took me inland, was rather obnoxious, feigning assistance in too helpful a manner, and although I don't want to accept favors from him, I will allow him to send important species on to you.

Many unusual events are happening that would make good newspaper accounts, but I hardly dare write them up. The weather has been good and I have been so active; I am hungry all the time. I keep hunting for the "unseen" through the rain forests and waterways but am overwhelmed by the gargantuan sizes of all species of flora and fauna. One cannot help but witness the coexistence of beauty with despair in the ecological struggle of procreation in this environment. I am sending my camera cartridges to Miss Northrup in hopes they are being developed in Milwaukee.

Yours Truly,

H. B. M.

*Article to the* Milwaukee Sentinel

## Trains—The Unique Cog Rail

The train at Santos is a remarkable feat of engineering. Actually it is two trains that run on a gravity road. The one going down the mountain pulls the other one up. They have not had many "runaways!" Two-thirds of the coffee from Brazil is carried over this road. The area is the dense tropical jungle type country that I have associated with in South America.

I had a good train from Petropolis (the Rio to São Paulo run). This is on a par with our trains with the same stifling curtains, upholstered seats and narrow aisles and only one berth to a section. But not to have a passenger above me was luxury. The linen sheets were clean and I was grateful for a woolen blanket. The nights are extremely cold. English customs are prevalent in the hotels of South America, where for example, one sees boots lined up in the hallways waiting for a bootblack. On the trains the custom is different. I found out when I did what Lloyd Morgan did when he stayed at the house of Robert Louis Stevenson. I put my dusty Oxfords outside my bunk with all the confidence of the English. Morgan added that "overwhelmed with such faith in providence, he was compelled to clean his own boots." I should have had the same compunction, for in the morning, I had no shoes. At first I was not perturbed for I concluded that it was too early and perhaps they were being cleaned. Rest assured, they finally came back—from having been delivered first to a gentleman!

The railroads are largely in the hands of English syndicates, and the laws of the country are such that when the earnings of the road reach a certain amount the tax shall be proportionately increased, but the income never reaches the desired amount, and yet the roads are said to pay 80 percent on the investment because of this clever evading of the law.

The South American is not a financier, and when he has to deal with a keen foreigner he is outwitted at every point. The electric street railways are largely in the hands of North Americans and must be paying investment. The roads at São Paulo are as well and perhaps better equipped and managed than our own. James Mitchell, who is in charge, possesses a wonderful combination of force, energy and power managing big enterprises, but is extremely jealous of any foreigner who is competitive which requires the most careful tact and handling to induce Mitchell to accept a secondary place.

*One of the trains taken through the mountains by H. B. Merrill.*
—Photo by Gathmann Hermanos, Caracas, 1900

This trip has been a diversionary novelty in the sense of transportation. I have been on ships, river boats, horses and carriages and trains. The cog train has been the most novel and exciting means of reaching high altitudes. When it starts, one hears gears gnashing, metal striking metal and the engine panting like a jungle cat. It is an experience I will never forget.

H. B. M.

Dear Nathan,

The cog train to São Paulo is one of the most remarkable feats of engineering in the world, considering it was constructed in a densely forested area. The roads are well managed by an Englishman, James Mitchell. The gravity system of the train going down the mountain pulls the train going up, with engines and pulleys. There have been few runaways. Two-thirds of all the coffee of Brazil goes over this road and stops are made along the way to pick up wet laundry which is taken to São Paulo to dry. Constant rain in the lower region makes it impossible for people to dry their clothes. Cisterns in those areas hold the water and release it gradually so as to prevent soil erosion.

Mr. Frank Carpenter, an American teacher in our company, showed us the house on Alto Sierra where the sailors are put up when they are in port. The Yellow Jack is too fatal to allow them to stay down in the port city even overnight. Every measure is being taken to control the disease. Seeing a jaguar leap from the siding on one of our stops wasn't as frightening as the prospect of the debilitating yellow fever. No mishaps thus far.

Love,

Hattie

## Journey Through South America by Enterprising Milwaukee Woman

*Article to the* Milwaukee Sentinel

## Boots Are Objects of Curiosity

I am both amused and frustrated when my shoes disappeared on trains and boats, where they are whisked away overnight to bootblacks who think they are men's footgear. On a train going to São Paulo, my shoes were missing in the morning. Since I had not packed a second pair and we were obliged to leave the sleeping car early, I tried my feeble Portuguese to summon the porter. I got no further than "onde estao" when he said "sí—agua" and started away to get some. My boots were recovered at last from the opposite end of the car! Most women, who wear tight high-heeled slippers, find the low heels and broad soles on my shoes objects of curiosity. It is possible that I could pay expenses by exhibiting them from town to town!

The porter returned with a pitcher of water and an enormous towel remarking that, "light skinned people have a strange fetish about washing." After inquiring where the lavatoria is, a lady in the next berth motioned toward two swinging doors marked "Señors and Señoras." I opened the door for ladies to find the occupant a man in a union suit! When I informed the porter, he showed me that the two doors led to one lava bowl and toilet and said he would hang a card on one door when the room was vacant.

I was finally let into the ladies' side which reeked with a combination of cigar smoke and Bay Rum. The motion of the train kept opening the door and every time we rounded a bend my morning ablutions were in full view of those waiting to get in. I offered the porter a milreis to hold the door closed for me but he didn't comprehend and the door swung open again. I gave him another milreis, and he took his position at the door like a sentry! I was relieved to see the red dirt that had been inches thick on all my exposed surfaces swirling down the drain. South American women, I was told, never bathe on board trains or boats.

*Article to the* Milwaukee Sentinel

# Officials for the World's Fair in St. Louis

Through the kind auspices of the Consul General Seeger, I am a guest aboard the same train that is carrying a part of the St. Louis Fair committee consisting of Colonel Bryan, minister to Brazil; Mr. Buchanan, ex-minister to Argentina; Mr. Bicknell of the United States Department of Agriculture and at least a dozen others. They have invited me to join them (as a representative of my country) at their breakfast and dinner meetings. Other women in my coach were curious as to what possible interest I could have in men's business or in the opinions I might offer them. In order to avoid the dilemma of being one woman among my countrymen, I pretended to be a stranger to them, explaining that I would call on them only in case of great need.

The second morning out, I had just ordered roll and coffee in the ladies' coach when Mr. Johnson, Mr. Bryan's secretary, with whom I have traveled from New York to Rio de Janeiro, came to tell me that the committee insisted that I join them for breakfast. I consented to let him escort me. It was one sure way of having a decent meal, I thought.

The dining room was a mere shed with long board tables like those in German beer gardens. The car was illuminated with flickering acetylene torches that gave off a lot of black smoke. In the early morning dark, the scene was eerie, and we all reminisced about the doughnuts and pies we used to have with coffee on the Pullmans back home. My countrymen continue to now ignore Portuguese prejudices against women traveling alone and have included me in everything on the rest of the trip. Judging from what they tell me, South America will have some new and exciting exhibits at St. Louis. The gentlemen were interested that I knew Theodore G. Sullivan, whom I met at Cornell. He is general manager of Standard Oil of South America, perhaps the best North American business in the country.

*In 1900, electrical power was showcased in dramatic displays at exhibitions such as this generator at the world's fair which blazed with lights. This original lithograph is from H. B. Merrill's collection and said to be from a drawing done for the St. Louis Fair. It is signed Graham.*

*Article to the* Milwaukee Sentinel

# South American Oil Industry

*Ships loading pitch from wharf at Pitch Lake. Few, if any women, had visited the lake at the time this photo was taken.*
—H. B. Merrill photo

    Kerosene is an absolute necessity in South America. There is little native coal. Wood is sold everywhere but kerosene is the cheapest and most convenient fuel. The large cans in which oil is shipped are used as wash boilers and roofs for shacks. There is talk of sending oil in tanks. I hope it is no longer put aboard riverboats as the oil permeates even the food. Standard Oil has a monopoly on lubricating oils. The company supports an army of people: engineers, agents, superintendents and lawyers.

    A Belgian engineer told me that he had discovered good illuminating oil in western Argentina and was going home to get capital to invest in the market. The Standard Oil legal counsel claimed that the kerosene in that particular area of Argentina was, as he could tell from the nature of the geological formation, inferior.

    An Argentine company invested a half million dollars into a field in Mendoza and another in Jujuy, which was located in a Jurassic formation and not as old as the Silurian deposits (I went out to both sites and took photos). That puts it earlier than our Pennsylvania oil region. The depth of the bituminous strata varies from a 30- to 50-degree dip and is separated by loamy deposits. The shale is so impregnated that it burns easily. When wells are disturbed by earthquakes, they flow spontaneously. In fact, in some places it has run off as waste into pure asphalt beds. One covers five acres averaging sixteen feet deep, bubbling up from the black spring. It could be considered the fertile crescent of the Americas.

    Petroleum from the Mendoza fields tested 25 percent kerosene when at 150 degrees while Russian petroleum fields yield 28 percent and samples from Pennsylvania tested 75 percent. However, Mendoza protects its export of oil with a heavy duty in retaliation to the United States tariff which has in the past favored Australian wool over Argentinean wool. I felt pleased when Mr. Bicknell (a company legal counsel) said he didn't think even Ida Tarbell had actually visited the asphalt "lakes" as I had.

    H. B. M.

*Article to the* Milwaukee Sentinel

# Rubber Trees and Plantations

To reach the interior villages I have traveled the tributaries of the Paraguay, Paraná, La Plata and Amazon river systems. The most exhausting were the myriad of tributaries with fields of grassy reeds that parted and closed as our riverboat "mowed" its way beyond Porto de Moz and on to Manaus. On the trip we discussed the fact Manaus was built by the "rubber barons." One German gentleman aboard said he thought the bottom might fall out of the market because of the slowing of production rate in South America.

I am fascinated with the towering *Europhorbiaceae* (rubber) trees. They grow upward to 60-feet in height with trunks 8-feet in circumference. So-called wild rubber is becoming scarce here and most of it is being shipped from the Pará province of Brazil. The British have invested millions in sterling on rubber companies in Ceylon and Malaya where there are workers to harvest enough to keep up with the demand. Ironically, the trees developed in the Pará province of Brazil were transplanted to those Asian plantations.

Fig. 1.—*Hevea brasiliensis.*

Although some plantations cultivating *Hevea brasiliensis* plants were developed to provide a source of national trade and ultimately local trade, for the diverse and scattered tribes, it has not been uncommon to find that, after working the land, some natives have set fire and burned entire plantations. One crude rubber plantation was surrounded by a large group of Guarani Indians who bound and gagged the owner, tying him to his running horse. A leader of the group, recalling that the Englishman he had been working for was fluent in the Guarani tongue and had overheard him planning the reprisal, realized he would be severely punished. He then talked his tribe into releasing the owner but not until much damage had been done to a vast area of cultivated land. In retaliation the British attacked the fort at La Guaira, Venezuela.

There have been tremendous investments made in over one million square miles in Brazil alone and the so-called "rubber barons" became incredibly rich in a short span of time. I am amazed at the great numbers of investments from rubber profits that were funding the development of cities. "Nitrate barons" profit as well.

H. B. M.

*Article to the* Milwaukee Sentinel

# A Gem—Set in the Prongs of the Amazon

Manaus is a prime example. It is uniquely situated 1,000 miles from the ocean in the upper Amazon region, surrounded by wilderness, high on a hill 1,065 feet above sea level. Some call it "Edwardo's Folly," after Edwardo Goncalves Rivero, a military engineer who envisioned a shining city—nonpareil. Foreign corporations are building quays and floating wharves because of constant changes in river levels there. Electricity has been installed in new government buildings and to run trams.

A wondrous theater, where operas and plays are performed by artists from around the world, opened just six years ago. For its construction, iron structural beams were shipped from Glasgow, Scotland, and then hauled up the hill. Italian artists were hired to decorate the interior, and the latest appointments were ordered from France. It eventually cost the investors 400,000 pounds—for culture. I find it astounding that great artists, etc. have found it appealing to perform here since getting to Manaus is hazardous and far from enticing or predictable. I considered it my most egregious folly to have even considered coming!

H. B. M.

*Entrance to Teatro Amazonas at Manaus, Brazil*
—*from H. B. Merrill collection*

*Report to the Milwaukee Public Museum*

# Culture of the Rain Forest Dwellers

From the Río de la Plata—the estuary is roughly 50 miles wide! We have come a long way from La Plata on the Paraná River system, stopping at Rosario and then on to Corrientes. From there to Asunción we are not far from the dread Gran Chaco region. It is said that the Tobas and Mattoccas are still cannibalistic, keeping shrunken heads of their enemy tribes. The captain said that several explorers who have gone into the forest there have never returned.

Just as with any other peoples, various tribes have different characteristics. Some are quite docile. One morning near a camp in this "Mesolithic Eden," I witnessed a most peaceful sight. A family of naked male and female natives were grooming one another's scalps and bodies, cracking the vermin found, between their teeth just like primates!

The Guarani people in Paraguay are of a pleasant disposition. Their features are refined rather like the Japanese and the women are very attractive until they age. They are worn thin from doing all the hard labor while the men lay in hammocks and chew on coca leaves which seems to put them in a debilitating stupor. Their teeth are even and sharp but darkened which dims their usually bright smiles.

Some of the men are not much taller than I. Following a small group into the forest (where it was cooler than near the river), I felt both welcomed and threatened by the cacophonous uproar of the howlers in the upper canopy together with the screeching bell-birds and startling flashes of macaw brilliance. One of the youngest men, proud of his accuracy with the curare blow dart, felled a parrot, cracking its spine with his teeth and took it back to camp. I refer to these settlements as camps because certain tribes move often to clear space for new sites. When they approach another group it is with much noise. Silent movement means stealth as in hunting or as a foe.

I admire the skin of most natives. It seems to be impervious to even the pernicious motuca fly which leaves a deep wound. But then, to iden-

—*Photo from H. B. Merrill collection*

tify their tribe, they pierce, puncture, paint and tattoo their flesh and insert all manner of objects into it. The Botocudo family men wear balsa disks as big as saucers in their ears and mouths! It is difficult to determine their general health unless one were to live among them. They are mostly hunters and fishers, which is the mainstay of their diet. Large numbers of tribes have died off but not, as some would have us believe, from outside influences. They are very territorial, killing or enslaving those who enter their camps.

It appears there is some tribal envy. When those from the interior venture out to drier land and see their enemies clearing large areas through burning and cutting, they threaten them. The fierce Mbayas force themselves on the agriculturally inclined Guanas, acting as "guards of the crops" in return for food and shelter. If Guanas resist, they are driven out or killed. When the land is exhausted from planting, the tribes move on to another area. Cultivating manioc results in a more predictable food source than hunting wild animals in dense forests. The manioc looks like our yams. I was going to try some that was being prepared by the women who chewed the tubers and then spit out the fibers into a communal bowl where it was left to ferment. I did not wait to see how it tasted.

The eating habits of the rain forest dwellers have probably not changed in a thousand years. There are many species of monkeys, large birds like the Jabiru, huge rodents, many marsupials and placentals, and most all of these are hunted for meat. The peccary, which is a wild pig, is very much favored but I can't abide the thought of tapir meat. The odor of the animal itself is most peculiar. Of course, there are many forest birds but the quetzal seems to be protected—perhaps by legend—from harm. I saw Indians making a meal from large lizards which they claim tastes like chicken. After reading R. Kipling's poem, about the Armadillo, I was quite devastated to come upon one, rolled into a ball and being roasted for a feast. It is said to taste like suckling pig. I was told that there are nine genera and 20 species of this armored *Endentata*. The Indians certainly don't lack for food with protein benefits. I have seen them chop down a Chambira tree that had telltale maggot holes and scoop up hands full of fat white grubs for lunch!

H. B. M.

*Native Paraguayans.*
—*H. B. Merrill photo*

*Guarani Indian boy.*
—*H. B. Merrill photo*

*Letter to the* Milwaukee Sentinel

*M. L. Hartridge sketch done in Dana Gibson style.*

# The Big Fishing Expedition

Back on the riverboat, I was followed about by a cheerful Guarani boy. He was fascinated with my note-taking but absolutely puzzled by my "fishing nets." For that matter, so were the passengers. Undoubtedly the men were familiar with the giant catfish Piraiba that can weigh over 300 pounds and measure 6-feet long. The Pirarucu is often 10-feet long. The Indians cut them in strips and dry them on scaffolds. One morning when I left the boat to go to Lake San Bernadino (Ypacarai), the Indian boy came to help me carry the equipment needed to catch the 1.1 millimeter elusive *Bunops* "animal." As I walked down the gangway, holding aloft Dr. Birge's fine silk nets like banners, and the Guarani boy tagging behind with a box of little bottles, I think the boat might have keeled over from the weight of every man on board leaning over the starboard side—gawking with curiosity over what could possibly be my catch of the day!

Sincerely,

H. B. Merrill

*Article to the* Milwaukee Sentinel *published in 1903 and portions of a letter to Dr. E. A. Birge*

## Steamer Trunks and Intriguing Passengers

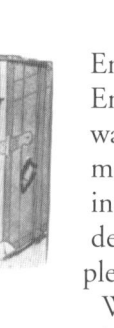

I am on board the *Magdalena,* an English royal mail steamer. The English woman in my cabin has a wardrobe steamer trunk with as many fashionable ensembles inside as the collage of foreign destination labels that completely plaster the outside.

We scarcely had time to freshen up when a steward rapped on the door announcing that dinner would be at seven o'clock. My cabin mate assured me it meant dinner jacket, tuxedo and vest for men and for women anything from a silk blouse and walking skirt to a decollete gown. It is cold on deck at night and my warm suit is most comfortable.

Among the interesting passengers aboard is the stunning, fluttering French "butterfly," the famous actress Réjane. The mere lifting of her skirts above the ankle caused a stir equal to the raising of the theater curtain on an opening night! She was very amiable and agreed to sign autographs during the evening.

Pat Sheedy, the notorious gambler, is on board dressed as a conservative businessman. He claims to have opened a jewelry business with his collection of precious gems from South America. He is an agent for Pierpont Morgan and is trying to solve the mystery of the missing Gainsborough, etc. Although rumored that his business ventures have actually failed, we noted he exchanged many English pounds and Argentine pesos during his stay.

Captain Strong and May Yohe (under the name of Mr. & Mrs. Atkinson) are quietly, but completely ignored. There is much entertainment with language games after dinner. One bright young Oxford man was truly the entertainment of the evening when he suggested playing word games. Samuel Pepys would have thoroughly approved, and I must say the two of us enlivened those gathered with our quick

*Réjane the famous French actress.*
—*Photo by Le Figaro in Paris, 1900*

44

repartee. Our exchanges of quotations, though completely impromptu, were heartily applauded. Except for that it was a well managed, typical English social evening, from the elegant main courses at dinner to the Nesselrode pudding. Count Nesselrode might have added a few extra chestnuts and rum, but as Pepys would say, "Strange how a good dinner and feasting reconciles everybody. (He might have included games). And so to bed," as he would say. *[to Birge]* Wish you had been here.

    H. B. M.

*[Merrill wrote similar letters to both her brother and Dr. Birge attempting to assuage any concerns they might have had regarding her safety. Actually, she was in constant danger from many sources.]*

## R.M.S. Magdalena

Friday, August 7, 1902

Dear Roger:

    You see I am not letting the grass grow under my feet. I have met some Americans by the name of Sullivan, very pleasant people. He is a graduate of Cornell and his daughter who has been two years at Wellesley College, returns in April. Mr. Sullivan travels for the Standard Oil Company, has been here half a dozen times so knows the people. They invited me to join in their company to Buenos Aires and as travelling alone is very difficult I was most happy to do so. I expect to go back to São Paulo and Río if they get rain so the fever abates. I am in no danger and I expect the tales you hear are worse than we know. The disease is very fatal with the English and Americans, but they steep themselves in whiskey and are out in the night air. It will be a very hard trip from Río to Pará but the American consul, Mr. Seeger, who knows Mr. Payne, has told me I can go and I think will make it possible.

I have been most fortunate. My good luck astonishes me every day. I have not travelled one step alone, and have not drawn one cent of money, partly because Mr. Purchase and Mr. Sullivan have advanced and partly because people are so glad to see Americans they will not take pay for board. Of course I try to make it right so I am not a burden financially. Delays, etc., make travelling expensive. The *Magdalena* was behind time at Santos and I paid four dollars to stay overnight. I am invited to stay a few days at the home of the director of the Botanical Garden when I return to São Paulo and I expect to go into the interior from there.

I was invited to the home of the sister of Santos Dumont (the airship man at São Paulo). The Dumont's are the wealthiest coffee owners in Brazil. I was taken all over the house from cellar to garret because she said her sister had so much wished to see the interior of an American home when in New York, but had failed. The most striking thing in the house was the solid silver washbowls and pitchers in all the bedrooms. Pat Sheedy (N.Y.) is on board and I shall have some interesting tales to tell of him. Of course, I am on my way to Buenos Aires.

My fishing seems to be most successful. I have found the very man who has what little collecting has been done here.

I have found the most beautiful old Portuguese silver and mahogany, also opals, amethysts, and topazes.

Tell Mrs. Hare I enjoy the Brazilian food very much.

Your Hattie Bell


*Article to the* Milwaukee Sentinel

# Currency "Control"

I am perplexed in planning my journey as to the best way of carrying money. Even my friends who have made European trips give varied advice, so at last, in order to profit as much as possible by their experience, I decided to take both a letter of credit (on the London and Brazilian bank) and American Express company checks. I did not take any gold, though advised to do so. American Express company checks, desirable as they are for a European trip, are little known in South America, and therefore, not good. All my orders are cashed by the London and Brazilian bank, but had I offered them in smaller towns, I am told they would not have been accepted. The London and Brazilian bank is courteous to me, and a letter of credit on them is good on the east coast, but the London and River Plate bank does business everywhere in South America.

English gold is always acceptable, so one should carry a certain amount of gold, or better, paper notes on the Bank of England. The paper notes are satisfactory, but as they may be endorsed without identification, are not as safe as a letter of credit. I soon learned how to carry gold, and often awaken so strangled by the weight of it around my neck at night, that I sometimes think if I escape yellow fever, bubonic plague, smallpox, and leprosy, the coroner's verdict will be "suicide by hanging."

The paper money, while largely printed in the United States, is of inferior quality and of such small denominations that an enormous purse is necessary in which to carry it. Coins are equally clumsy. A Brazilian five-cent piece is as large as our half-dollar, although the newest coin is a pretty piece which looks like our five-cent piece in that it has the goddess of liberty on one side, but has the southern cross on the reverse side. The unit in Brazil is the milreis, worth almost 25 cents today in our money.

In most other countries the peso is the unit. It differs widely in value in different countries.

South America is in advance of us in that the metric system of weights and measures is in common use, but it takes some calculating for a North American to buy cloth by the meter, and alcohol by the liter. The money, too, is a great nuisance because it is different in every country, and its gold value often changes every half hour. It is not difficult to understand, however, because a decimal system is usually used.

Am not seasick this trip.

It is cold, cold.

H. B. M.

Buenos Aires
August 11, 1902

Dear Dr. Birge:

I was quite successful in my fishing at São Paulo. There is a fine museum there and the director, Herr Von Ihering, is the one who sent my mud to Sars. He had all of Sars' papers and loaned them to me and stated that I had collected samples from further north compared with Sars' Macrothricidae. The forms he raised were mostly mud forms. The material looks quite dry. I asked Von Ihering to get some fresh mud samples from the same places he sent Sars to collect, as my boat is leaving to go back to São Paulo. I copied the dates and locations and will enclose them later. There will be three papers. Richard has done some work from dried material here, and I plan to continue on because I know you want to establish populations.

H. B. M.

*Birge saved the samples some of which are extant from Merrill's research in São Paulo. They yielded* Macrothrix, Squamosa *and* Ceriodaphnia rigaudi. *See Part V: scientific charts starting on page 186.*

Phoenix Hotel, San Martin, Argentina
Tuesday, August 26, 1902

Dear Roger:

    I suppose you think my letters too few, but I am going every minute and only hear that mail is going out a short time before the boat sails. I start Thursday for Asunción and the Iguazú Falls. People tell me that not more than a half-dozen white women have made the trek to see them. Miss Sullivan, a Wellesley girl, daughter of Theodore Sullivan, general manager of Standard Oil in South America, has accompanied me on some of the journey but I don't think will go further from here. The falls are said to be larger and more magnificent than Niagara but hard to reach.

    Mr. Lord, the minister here, and his assistant, Mr. Ames, have ignored my letters and requests for getting about. Mr. Sullivan (a nephew of "Sunset" Cox) wishes to make a formal complaint against them. You may tell Mr. Payne if you think it best. I am succeeding splendidly through English! Have met the Argentine minister to the United States during the Cleveland administration and he has obligingly given me letters to all of the governors of the places I am going. I could not go without them. One of the ministers, Señor Zaballos, who was minister during McKinley's time, was most accommodating and I must write to Professor Birge about him.

    Tell Mrs. Hare I am able to converse in English and have people to talk with.

    With love,

    H. B. M.

*Eucalyptus Avenue leading to La Plata Museum.*
—*H. B. Merrill photo*

*Merrill wrote from San Martin to Dr. Birge.*

Tuesday, August 26, 1902

Dear Dr. Birge:

    On Thursday I start for Asunción and the Iguazú Falls. I have heard that only a few white women have ever seen them. I can't send my photographs from here but am getting a considerable amount of zoological material.
    The South Americans pride themselves on the beauty of their scenery. They say the North Americans are practical, they can build bridges and railroads and great cities, but South America has preserved nature, and they point to their great mountains and plains and rivers and ask what we have to compare with them. They even claim to have waterfalls more wonderful than Niagara, higher and larger, in every way. These falls are situated in a branch of the Paraná River between Brazil and Argentina. The journey to them is a distance of 1,000 or 1,200 miles, but a North American cannot let the Niagara challenge pass unheeded, so we set out glad to journey up one of the great rivers of South America.

    Yours truly,

    H. B. M.

*La Plata Museum.* —*H. B. Merrill photo*

*Report to the* Milwaukee Public Museum

# Nightmare Trek to Iguazú Falls

From Corrientes to Posadas we saw many caimans that, at first glance, looked like big black logs strewn along the shores. Lethargic as they seemed, these particular crocodiles stopped dozing and rushed the boat. Further up the river the banks become higher. Bright green bamboo shoots rise tall above all other foliage with clumps at their tops resembling the Prince of Wale's crest. All of the Lapacha trees are profusely covered with brilliant cerise and yellow bell blossoms. This glorious color, reflected in the early evening calm of the water, seems less like spring and more like Indian summer back home. Sunsets here are ethereal. The sun hangs in the sky, a great red ball never seeming to set, just slipping away near the horizon like a phantom enveloping everything in misty opalescent tints.

When night finally falls it's with a sudden crescendo and, though there is a generator on the riverboat, the captain advised lights out so as not to attract buzzing, stinging clouds of insects. He is particularly concerned about mosquitos and asked me about the anopheles. I told him that I wanted to write notes in my journal and he allowed me a small kerosene lantern next to my small bunk. We have netting to wrap around ourselves against a cloud of black flies that began an attack about sunset and have bitten some in our party. Although I am not molested I can't sleep. There is an incessant sizzling, croaking and growling of nocturnal creatures. Except for an occasional flicker from the fire or torch of a native camp, the night is suffocatingly black.

As much as I wanted to see a jaguar I was glad to be anchored off the landing. I did have the sense that I was not alone while writing. Every time I stopped scratching on my note pad I heard a "tick, tick." Each time I was still it grew louder and closer until it stood straight up in front of my lamp, a glowing, absinthe green, moving piece of orthopterous "vegetation." It was the tallest praying mantis I have ever seen, and I was torn between getting help to capture it as a specimen or swatting it with a rolled up paper. My camera was packed but the light for exposure was not adequate anyway. I decided to swat the thing, a problem without knocking the lantern over. Companions on board must have heard the commotion as I shoved the animal out the door and onto the deck. Deciding to keep the lantern lit, I got into my bunk and as I settled down, the "tick, tick" began again. There, with the light shining through its eerie green body, eyes glistening on its swiveling head, was my defiant tormentor. I was embarrassed to have to ask for help in such a situation but I had all I could take of the "Edgar Allen Poe nightmare." We finished him off thoroughly—denying the female of his own species the pleasure of devouring him!

As hideous an entomological specimen as the giant mantis was, I can imagine how intriguing it would be for the boys in our classes at the museum to see it in action! I am collecting some items of interest for your anthropological studies. When I get to the interior where I can observe the cultures of native tribes, I will send notes on to you. It has been difficult to keep both handwriting and correspondence steady. On to Iguazú Falls.

H. B. M.

*Los Mantis danza—the dancing mantis.*
*M. L. Hartridge drawing.*

*Article to the* Milwaukee Sentinel

# Trek to Iguazú Falls

The rain forests are so thick and wild that for hundreds of miles we scarcely see the banks, which are low and marshy, with great bare stretches of sand that remind one of the New Jersey seacoast.

The river is muddy, much worse it seems to me, than either the Mississippi or Missouri. Our boat is stuck on sandbars several times but we are not seriously delayed. The mud settles on sticks and debris forming an island and during high water may become detached again and float down the river. These so-called floating islands sound much better than they look. As we near Corrientes, the river is dotted with islands, but they are usually flat patches of sand with, perhaps, a few water weeds, or there may be stunted trees and shrubs, which as we go further on, are completely shrouded in a grayish, green parasitic moss or vine.

The rapids here are a source of interest to the South Americans, but seem tame to us, for there is little to see, although the current is swift.

The birds are numerous and tame. They hardly trouble to fly away as the boat approaches the shore, but even at that short distance it is hard to recognize the unfamiliar forms. From Corrientes to Asunción crocodiles are common. Usually they are too sluggish to move unless the boat passes near shore. They are strewn along the shore like great black logs.

Our party to Iguazú consisted of two guides, a botanist from Sweden, an entomologist and I was the only woman since Miss Sullivan did not continue the trek inland. Just before we reached the Iguazú River, the depth registered 1,000 feet and carried a tremendous volume of water. On both sides of us beautiful waterfalls poured over black basalt rocks surrounded by bright foliage. This view alone amply repays the hardship of the long journey. In the hazy sunset we anchored for the night knowing the greater falls would be seen tomorrow.

We started out on horseback through dense foliage about 20 miles from the falls whose roar from the cataracts echoed through the forest. Moisture spilled from every leaf end under an eerie green canopy and there was such a tangle of growth as to cause all forms of life to develop into eccentric forms in the struggle for survival.

We continued on foot through barbed barricades of epiphytes and parasites as the path narrowed. Underfoot, the *Selaginella*, liverworts and mosses were equally precarious, and I sunk up to my boot tops every step of the way. Some tree trunks with six-inch thorns sticking out in bunches grabbed my clothing, and by the time we got to a clearing, I felt I had awakened from a nightmare. At last we reached our destination. Of all the descendants of primordial species I have confronted, at least I have been spared the vengeance of fire ants. Our guide tells of a party of British men who were found totally devastated by them—down to their leather shoes, clothing and bones which were scattered in bits along a trail!

At the confluence of the Iguazú and Paraná rivers, 275 cataracts define the borders of Argentina, Brazil and Paraguay. The position of the falls is similar to Niagara but more than two miles in the diameter of the semicircle. A great vertical cliff divides the main body of water that projects cascades of great volumes of water that thunder over the rocks in the midst of the absolutely vast wilderness that surrounds the falls. It was worth traveling 1,200 miles to see.

The night before we left we glimpsed it during a tropical storm when the heavens were ablaze with sheets of light. The sky seemed to break open and the rain fell in torrents. Combined with the thunder of the falls, it was as though the voice that spoke to Job out of the whirlwind was saying: *"Where was thou when I laid the foundations of the earth? Declare if thou hast understanding. Who hath laid the measures thereof, if thou knowest? Or who hath stretched the line upon it? Whereupon are the foundations thereof fastened? Or who laid the cornerstone thereof; When the morning stars sang together, and all the sons of God shouted for joy?"*

*Report to the Milwaukee Public Museum and the* Milwaukee Sentinel

# Light in the Rain Forest

It is difficult to describe the kind of light one experiences when deep in the rain forests. It is almost impossible to see the sky even through an occasional opening as the leaves above are so interwoven. Yet, it is not at all dark or gloomy.

The story was told by Walter Damrosch that when he was in the Adirondacks with one of his children they saw a beautiful sunset one evening. The child responded, "Isn't this just like the second act of Siegfried?" To which his father, surprised that he was so impressed with the artificial, replied, "No, but the second act of Siegfried is a little bit like this."

In order to get some idea of the unusual effect, I must appeal to the artificial and say it is like the green lights projected on tableaus in the theater. Even with all of this wealth of light and life, it is still not as impressive to me as our Northern primitive white pine forests.

H. B. M.

*Article to the* Milwaukee Sentinel

# The Brown Amazon and Tributaries

    To reach some of the interior villages I have traveled on riverboats, the tributaries of the Amazon—Paraguay, Paraná and La Plata systems. An Austrian named Nicholas Mihanovich has a complete monopoly on most of the boat lines. He has bought out most of his competitors and can set the rates as he wishes. He charges double the rate going back down the rivers as he does going up. Freight rates are higher from Buenos Aires to Asunción than they are from Buenos Aires to New York. Mr. Mihanovich has a genius for organization and a mania for seeing his name displayed. Every dish, every article of furniture, towel and windowpane on his boats bears his name. The most conspicuous flag on all the docks has a red ground with the large black letter "M," which floats over the offices of his agents.

    A good line of flatboats, like our Mississippi riverboats, would be of light draft—the better to maneuver the treacherous La Plata River. Our boat has been stuck on either sandbars or floating islands several times. Mud settles on debris that tumbles into the water from high bluffs at high water. Grayish green parasitic vines cling to fallen treetops that ride on top the water. A disturbed water boa just unraveled itself from a woven mass of vines suspended over our heads, then dropped into the river, dissolving in the murky ooze.

    On the Paraná and Paraguay Rivers, the rapids are swift in some sections and swirl us about in a game of plunge and spin about at which the pilot is a master. We are on our way to Asunción.

    H. B. M.

*Article to the* Milwaukee Sentinel

# On to Asunción Unaccompanied by Men

I am accompanied to Asunción by a young Wellesley college woman and our reception at the Franco-Argentine hotel is hardly what may be called cordial. We have been directed in the hotel by some English people, but arrive alone. The proprietor, when we ask for lodging, says, "And your family?" We try to explain but he fails to understand and still demands the needful family. At last he goes to the door and looks out. His manner is perfectly polite but it is utterly impossible to comprehend our position. While foreigners disapprove of our travelling alone they seem to feel that since we are alone they must assist us in every way possible. A German gentleman speaks to the custom house official and our baggage is not disturbed. We find he has also requested attention for us at some of the towns we visit, and he does all this for us, although we have hardly exchanged a dozen words on the boat, and never see him again after landing in Asunción.

Asunción is one of the oldest and quaintest of South American cities. It is situated on the east bank of the Paraguay River, which is low and marshy. Its bright colored low buildings, nestling among the green foliage, and its country houses, with their wide verandas and beautiful gardens full of tropical vegetation, are the most attractive in South America. Asunción is a queer combination of decay and growth. Many of the structures that López built are falling into ruin. One good building with a handsome dome was intended as his tomb, but by the irony of fate, he is buried in the forest where he fell, and the building is in ruins. A few of the main streets are paved with cobblestones, but the roads are bad, especially in the country where the sand is deep. Just across the river is the Chaco. This region is an anomaly, for it lies just across from a city of 50,000 people and yet is occupied by Indians so savage that white men never visit it alone. Many have gone in and never returned.

H. B. M.

*Women of Asunción grinding corn.*
—*H. B. Merrill photo*

*A native conveyance from Asunción.*
—*H. B. Merrill photo*

August, 1902

Dear Dr. Birge:

Señor Zaballos, Argentine Minister, has been very kind. Mr. Vilas must know him. When I return, I should like to ask Mr. Vilas to write Señor Zaballos thanking him for me in the name of the University. He gave me an introduction to the Governor Lanusse of Misiónes, and I was royally entertained there. I have eight letters of introduction so expect a good time. It is hot today and I have on a cotton dress, the first I have worn as I have been frozen most of the time. Had flannel nightgowns made in B.A.

The falls of Iguazú were most wonderful. This journey to the interior was better than Paraguay. I am very fortunate or your good wishes have saved me, for the insects have not bitten me or it hasn't poisoned me. Near Iguazú, there were little black flies that poisoned some of the party terribly but did not touch me. Tell Professor Owen I have not needed my insect powder since the boat. Also tell him I have found some interesting Spanish literature and while I was in the La Plata region, I got a book by William Henry Hudson, a Scottish naturalist, called "The Naturalist on La Plata". I'm also sending some excellent stories on gauchos entitled "El Ombu", just published this year.

Argentina is said to have no native trees. If this is true, the foreign trees have been successfully introduced. The eucalyptus, or Australian gum tree, is a favorite because of its rapid growth, but it is ugly as it gets older. The bark peels and hangs in rags about the trunk, leaving the white cambiums exposed like bleached bones. The Ombu, one of the few large native species of Brazil, is really an herb. It affords the only shade on the pampas.

I am not forgetting you and have some things I think will interest you.

Yours,

H. B. M.

*A rare look at Argentine gauchos.*

*Gaucho using a bola to lasso cattle.*

*Article to the* Milwaukee Sentinel *and a Report to History Departments at Universities and Organizations With Which Merrill Was Associated*

# History of Paraguay

From Iguazú Falls our journey continued from Misiónes and Paraná into Paraguay, which both because of its history and its distance from the seacoast, is one of the most primitive countries of South America.

It is in measure impossible to understand the Paraguay of today without some knowledge of her history and the history of Paraguay is so little known to English readers that a brief resume of it may be of interest.

John Sebastian Cabot was commissioned by the Spanish to explore a southwest passage to the Pacific where he ended up discovering Paraguay and other interior countries in 1530. It was then occupied by a tribe of Indians known as the Guarani. The tribe originally extended from the Amazon to Patagonia. The remnant now left is found mostly in Paraguay. Soon after its settlement the Jesuit missionaries obtained exclusive dominion over the land and for more than 200 years ruled absolutely. No foreign influences were permitted, except what came through Jesuit channels, the Spanish language was forbidden.

A Guarani grammar text and a number of works in the Guarani language were published. Most of these books have now found their way into European libraries.

*Coolies hut in Paraguay's interior.*

—*H. B. Merrill photo*

# A Jesuit Stronghold

When I was at Asunción I was told that it had been a Jesuit stronghold since 1557 and was a territory of Misiónes and most of Paraguay. The docile Guarani tribes were easily led and they actually provided the labor that built most of the missions. One thing commending the Jesuits was that they built large compounds in which to house and teach the Indians how to weave items that they could trade or sell. They were also excellent at wood carving and some of their work can be seen today. They were subservient supplicants attending mass every day which lent some order to their daily lives. Then the Spanish explorers came looking for wealth along with Brazilian slave hunters. By 1767 the Jesuits were expelled from the country by a Spanish viceroy who thought the Jesuits had too much power over the natives.

I believe the Jesuit order had a good influence on the Indians in the beginning but they could not exist in the "routine" of the European mind concept of order. Without direction and instruction from the church they were lost. When left on their own, the Indians reverted to their original superstitions and in many cases, savage behavior. It is thought that there was more pageantry than thought processes in teaching the scriptures to people of a primitive background. When the leaders left, the missions decayed along with order and productivity. The well planned center plazas surrounded by workshops and provisions for the natives and even the churches were soon shrouded in jungle growth. There are still several little "stations" here and there for the Virgin Mary and many saints that are in pitiful condition.

One learns during travels from town to town that where priests still have a following, there is competition among them for which sanctuaries have the more important or authentic relics of the sacred departed. It is astounding to a North American that some priests in these countries have children of their own. One priest, living near Lake Ypacary, is educating his son for the priesthood. With the expulsion of Jesuits, Franciscans and Dominicans took their places in most of the countries. It is said that with all of the instructions given the natives, few of them take vows seriously and 90 percent of the children are (in the terms of the church) illegitimate. Even some bishops are not celibate.

## The Church and the Missionaries

Several years ago, a Protestant mission, the Argentine Methodist, was established by a Reverend Dr. Wood. It was the first time that the history of Christianity was taught in an all-inclusive manner rather than as a mysterious cult. The schools grew to such an impressive size in Asunción that when President Sarmiento, an Argentine educator, learned of its success, he wanted to model others after it.

There are many converts to the Protestant missions, Montevideo, Asunción and two theological schools in the German colony at San Bernadino. I see many attempts to prejudice the Indians against the obvious effectiveness of these educational centers. They and the Masonic Orders have done much to improve the general health and well-being of so many desperate people struggling to make a better existence for themselves, but the fear is that Protestant churches will compete with the Saints of Rome.

You can imagine that in a predominantly Spanish-speaking country, it has been difficult for

*The Catholic cathedral on the Plaza de Mayo, Buenos Aires.*

the English-speaking preachers to get across their good Christian works by any method other than being living examples of what they try to preach. Thousands of American dollars are put to use in these measures to start theological schools in Uruguay and Brazil. Offering the scriptures in translated form has resulted in the availability to all the people, the words of the Bible without interceders who it is said, "take native souls for ransom."

The Jesuits built their missions, and worked and taught as they did in other parts of Spanish America. The ruins of their buildings are found scattered all through Paraguay. The Guaranis proved docile pupils and readily adopted, superficially at least, the religion and civilization of their conquerors. But I have noted among most all of the native tribes that zoomorphism is intrinsic to their beliefs. The Jesuits were expelled by the Spanish government in 1767 and the Guaranis relapsed into anarchy and barbarism, which lasted until 1816, when Jose Gaspar de Francia was elected dictator for life, with absolute and unlimited power. He ruled until his death in 1840 and was followed by two other dictators: Carlos Antonio López and his son, General Francisco Solano López, who died in 1870.

## Career of Younger López

The career of the younger López, who ruled only eight years, was most remarkable. He was a man of much intellectual ability, great military genius, and vast ambition, but was unjust, and cruel; a man who, as his historians say, could only be understood in his proper setting, in his natural environment. It is said he was desirous of emulating in South America the career of Napoleon I in Europe. His ambition was seconded by an English woman, who shared his fortunes as his wife, but who was known as Mme. Eliza Lynch, for, during the dictatorship of Francisco, the marriage ceremony was discouraged and has not since been generally recognized in Paraguay. Mme. Lynch exerted extraordinary influence over López and the fortunes of Paraguay. A native book describes her as follows: "She was about 28 years old. The dark blue silken robe, closely clinging to her form, brought out the proud, flexible figure in strong relief. A cape of red silk, trimmed with snow white swansdown and covered with golden studs and black cords, and held by a heavy cord of gold, hung gracefully from her uncovered shoulders. Her luxuriant gold blond hair was arranged 'a la page' and held in a black silk net. A Hungarian barrette, adorned with heron plumes, fastened by a golden clasp, audaciously crowned her head, while out of the delicate white countenance looked a pair of dark blue eyes, shadowed by heavy lashes."

*Veranda of Eliza Lynch's country home. A very rare photo taken April 5, 1903.*
—*by H. B. Merrill*

*Ruins of one of the summer homes of Eliza Lynch.*
—*H. B. Merrill photo*

# López Plunged Country Into War
# Venezuelan Conflict

The opportunity soon came for López to further his ambitious schemes. Through what seems his unnecessary interference in the affairs of Uruguay, he plunged his country into a war against the combined forces of Brazil, Uruguay and Argentina, against troops numbering more than 15 times his own. The war lasted from 1865 to 1870 and we who smile at South American rebellions find that the generalship displayed by López must at least command our respect, although the atrocities practiced would hardly be believed possible in the nineteenth century. The result was inevitable, and at the close of the war, out of a population of 440,000, only 28,700 men remained and of these probably not half were able to bear arms. This is the desolation from which Paraguay is now trying to recover. The recovery has necessarily been slow and this, together with the fact, that until the middle of the nineteenth century the country was hermetically sealed to all foreign influences, makes it a fascinating country to the traveler. It seems to afford a glimpse into the sixteenth century.

H. B. M.

∞

Dear Cousin Nora,

I have written my general article on the contrasts of wealth and poverty in Buenos Aires to the *Milwaukee Sentinel* but simply touched upon the cathedrals and the church in general. You have always received my comments on the church (however contrary to your own) with patient restraint and intelligent interest.

To the *Milwaukee Sentinel* I described the Plaza Victoria and European-styled buildings such as the Casa Rosada and the Calle Florida complete with kiosks and people dressed in fashionable attire as emulating the Parisians. With the lights on at night, which seems to be when shoppers are out full force and carriages drive British style on the left side of the street, I have never seen a more festive sight. The most elite clientele dash from carriages to the elegant shops. Fabrics, furs and jewelry, sparkle throughout the inviting, well-lighted window displays. On certain other streets off the main boulevards, señoritas lean from unscreened windows or balconies on satin pillows looking out on the colorful parade below.

May Avenue is the first street I passed down when arriving, as it starts at the river and goes some miles to the commercial district of the city. It is the buildings themselves that fascinate one as much as the people. Most large cities have many Roman Catholic cathedrals built by the colonials—Spaniards and Italians. Many of the major buildings and particularly the cathedrals appear to have been constructed with much marble, then on closer

*Calle de Florida Street.*

scrutiny, you learn that they are surfaced with plaster that is skillfully painted to resemble beautiful Italian marble. The cathedral on the Plaza Victoria has a facade of Grecian architectural design and the interior holds treasures in gold, silver and gems; almost as grand as St. Peter's in Rome!

But the riches of the country housed in the churches don't seem to be distributed to benefit the ubiquitous natives who camp like serfs on the steps of the monuments to Christ. Slouched over in their ponchos, they beg for coins for lottery tickets or carfare to get back to the slums outside the cities. Many come in by oxcarts in the morning and sit in the plazas all day.

Forgive me dear Nora, but the Spanish occupation did not foster principles beyond fear and idolatry. Perhaps it was the only method to control the people and give orders but it did not teach them self-discipline. I am sure the Jesuits wanted leaders rather than sycophants to emerge. I did not see the renowned monstrance in the cathedral at Cuzco but was told it was cast from 50 pounds of Incan gold and encrusted with more than 1,000 jewels!

I must now find a typewriter and get my journal sent off by mail boat.

My love to you and the Emmons,

Hattie

# The Beef and Leather Industry

Dear Nate,

In talking with an investment broker at my hotel, I was reminded of a statement that Marie has made so often, "money talks." It is the same the world over. Down here there are differences in opportunities as widely disparate as any North American city. Here it is the military or government employees who are assured of at least a fair living wage. But it is the political officials who in most cases have amassed fortunes from mining, oil, rail systems, imports and various exports, particularly the beef and leather exports. Formal state occasions become elaborate and costly celebrations that would stagger the imagination of those who view our usual North American political functions. The wealthy women all have enough "handmaidens" to rival royalty. They are the royalty of the major cities in a sense. They have endless hours to spend in the shops or go off in their carriages to the opera or theater.

It seems to me that the educators—those in academic life—those dedicated to sciences and medical advancements, should be able to share some of these leisures, but that would become a tedious bore to those of us who find excitement and gratification in exploration, discovery or inventing that which could be useful to mankind of all persuasions.

*Argentine pampas cowhides drying, then to be shipped to Europe, mostly Spain, for tanning.*

*Merrill considered this hotel in Montevideo one of the best.*
—Photo H. B. Merrill collection.

Speaking of the beef and leather industry, it is mostly the women in many areas, who do the slaughtering! I have never seen such good specimens of beef cattle as have been shipped here—Durhams and Herefords. Most hides are shipped out of the country to be tanned, many to Spain. I am writing to the *Milwaukee Sentinel*, about the pampas where the cattle graze and when slaughtered, they used to leave carcasses lay out on the vast estancias to rot in the blazing sun. Large herds had been unfettered and left to roam to the extent that the meat is often too tough. Now, since the European demand for Argentine beef has become so much in demand, they have fenced in large ranges. It is like our great Western plains. The gauchos now brand their cattle. The sheep imported have such heavy wool they can scarcely stand from the weight of it.

As I mentioned in my article, the Chicago packing houses are said to be refusing Argentine beef. It is certainly plentiful here. I have had it served for breakfast, lunch, and dinner prepared in various ways and even beef bouillon soup precedes some courses. One way I have learned to enjoy it is in a stew with tomatoes and other vegetables, a sort of puchero.

Incidentally, in Paraguay we found some excellent hotels and restaurants in Asunción. Women are discouraged to go to them even for lunch, without a male escort, especially after 4 p.m. Edward Dent, an English gentleman I met on the boat, insisted on taking Theodore Sullivan's daughter and me to one of the cafes. We were treated to South American chocolate which is deliciously thicker and better than ours back home. We were told that on certain days they also have true North American ices and that it is one of the few places where women are welcome. This was delightful to hear and so we set out bright and early the next morning thinking breakfast hours were certainly safe. Miss Sullivan in her demure dimity with fichu and I in my sedate Eton jacket and skirt with boater, looked perfectly innocent, but neither of us was accompanied by Mr. Dent or any other man!

There seems to be a caste system for a woman's place in this country. They don't drive street cars, however the loading of cargo at the docks as well as unloading, is predominately

*Hotel de Paris at Posadas, Argentina. Merrill on right with Miss Sullivan and friend.*

—*Photo from H. B. Merrill collection.*

done by women. Yet, we seem to be considered nonfunctional without male assistance. To continue, Miss Sullivan and I seated ourselves at an outdoor cafe table and ordered that marvelous chocolate when suddenly, an audience began to gather three rows deep on all sides staring at us as though we were some rare species. A few went in to ask the proprietor if our men were inside? We learned that women may not be served even breakfast if on their own and that is a concession in many places in this country where women are welcome only as appendages to the males.

    Love to you and Marie

    Hattie

*Garden inside Hotel de Paris.*

# The Guest of Honor

*Report to the* Milwaukee Sentinel *and Note to Mrs. Hare*

Dear Mrs. Hare,

    At the Governor's Mansion in Misiónes, I was invited to dine with area dignitaries the other evening. There was much bowing and curtsying and kissing—something "foreign" to me. One gentleman who kissed my hand remarked in Spanish that it was, "A tasty little biscuit!" The hostess saw me looking at my hands which, browned and rough from exposure, must have felt coarse as crust. She leaned toward me explaining that all of the biscuits served at her affair were imported and therefore extra special—which somewhat reconciled me! I should add that this thoughtful wife of the Governor extended to me another courtesy which I will never forget.

    Since I was the guest of honor I arrived a little early, in what I considered my best traveling ensemble (boots and all). Typical of the elegant wealthy South American women, the Governor's wife and daughters were attired in silks, satins and opulent jewelry—some with stones as large as those seen in museums. Then, before the other guests gathered, the family quickly changed into their plainest cotton street clothes, tactfully making me seem less understated in costume! For these women, who pride themselves on owning gowns made in Paris, wearing anything less at a formal function was a considerate sacrifice. (To Mrs. Hare-They have insisted that I see some of the finer shops before I leave the country. When I do, I will send pictures or write about it.)

Godey's Lady's "Guru"

Hattie

*South American women attired in Parisian gowns. Art by Jacques Tissot, 1836-1902. French born and trained, Tissot achieved great acclaim as a contemporary artist.*

*Home of Señor Castelli, Governor of Misiónes.*
—*H. B. Merrill photo*

Dear Nora——Just a note

    One of the houses I was invited to dine at was that of a state official, Señor Castelli. He speaks no English but his wife and daughter speak beautifully. It was interesting to note that hostesses do not stand in receiving lines but are seated in a fixed place where guests present themselves to her. At several formal functions I have attended, perfume has been passed around by gloved attendants, much as we would pass candy favors.

    And now Nora, you are much the more fashionable than I will ever be, and I have tried to pay attention to what is being worn. The general readership will expect it. First: the Argentine women among all others, seem to be the most stylish and gracious in the manner of our own Southern women. They are a combination of several countries. The saying here is that "when Venus apportioned her gifts, she gave grace and elegance to the Spanish women, liveliness and savoir faire to the French, perfection of form and features to the Italian and to the Englishwoman, the complexion as clear as the morning…but the *porteña* women were endowed with the best of all charms in any country." You can bet that a man authored that legend. The typical Latino male psyche in much of South America is to dominate their womenfolk completely. I am sure I am considered a conundrum by men of such disposition. They, as General Francisco Solano López, like to emulate the dictator, Napoleon I. I am informed by a member of the consulate that Señor Solano López, son of the General who died in 1870, would like to have a meeting with me. I am told that he is a man of intellectual and leadership ability and with physical stature similar to the "Little Corsican!" I am intrigued…will report.

    Your undaunted,

    Hattie Bell

*Notes from Merrill's journal and a letter to Roger*

## Latino Beauties and Exuberant Equestrians at Their Exhilarating Best

Dear Roger,

Never have I seen such contrasts from my culture as out here. Fiesta time on the pampas and in the city is beyond imagination. Any kind of county fair, even in our western states, can't compare.

There are breeds of all horses. The wealthy gauchos and estancieros parading their Arabians, fitted in solid silver saddle trim, flashing in the sun, are magnificent. Elaborately appointed coaches, carrying dignitaries, roll along as exuberant, young caballeros, some riding barefoot and bareback, maneuver in and out of cheering throngs. Some gauchos from out on the pampas wear boots made of skins of freshly stripped colt's legs, pulled on before cured!

The women riding in decorated open carriages are elegant ornaments themselves. They dress in the flamboyant fiesta mode, covering their tan complexions with chalk white powder—a startling contrast to ruby lips and dark eyes accentuated with even darker mascara. Some comb their hair straight with spit curls at the sides and crown their heads with enormous jeweled "fan" combs, almost as big as their hand fans. The young girls toss paper favors or flowers to viewers along the streets, while bands with many brass instruments and guitars add a pulsating and rhythmic addition to an exciting and colorful event. It is a show of Latin blood with its most exhilarating equestrian beauty and skill—a complete antithesis to my middle western background. I'm afraid I am under a sort of southern culture spell.

In Paraguay, I met Señor Solano López, son of General Francisco López. He is charming and amusing, telling me that it is good that I have arrived between revolutions. He is handsome though short of stature, and the history of his family is fascinating, which will deserve more time to describe later.

As superintendent of schools, he escorted me about to the various districts, describing their educational systems. He is a proponent of Theodore Roosevelt and has published the President's speeches translated in Spanish for his countrymen, including the students in the schools. I wouldn't be at all surprised if he becomes dictator of Paraguay one day. He would be an excellent leader, having the stamina and patriotic zeal of our hero Roosevelt.

It would be of great benefit to our country if other heads of the South American countries were of the same opinion as Solano López. He is in full accord with Roosevelt's efforts to obtain the Republic of Panama for the United States. In return, it is the hope that the canal can be constructed. The French and British had undertaken the vast project and failed. One of the greatest hazards has been the insidious malaria and precarious terrain. Señor López asks, "why should our merchant ships have no alternative other than to sail around Tierra del Fuego to reach the Pacific?"

In closing, I must tell you about an amusing experience I had as woman. While at the University of Brazil, I had to walk from the laboratory to my quarters past a square near

*Grenadiers from the Plaza de Mayo.*

the barracks where the soldiers drill. They looked so splendid in their uniforms, but I was embarrassed when I had to cross by them near the plaza. The whole company stopped at attention, turning toward me, saluting as they addressed me in unison, "La chiquita Enriqueta!" The cheering and raising of their hats was embarrassing, but they did look smart and handsome in their uniforms. Although it was spring, it was a balmy summer afternoon and except for all of the exhilarating attention on me, I was giddily tempted to 'tour jetee' across the entire length of the plaza!

More about uniforms and Señor López in another letter.

Love,

Your Hattie "Belle"

## Observations on Homes and Social Customs

*In her articles written for the readers of the* Milwaukee Sentinel, *Merrill appealed to the women when she described the interiors of some of the homes of officials and their families who offered her accommodations. The social traditions of the mixed cultures were of equal interest. Her letters to Birge which mentioned her contacts in South America, gave little if any details of a domestic nature. The following are notes that Merrill made during her journey and sent on for publication.*

The women of South America are domestic. They have little opportunity for being anything else. I see some exquisite housekeeping, as insect pests are such that cleanliness is essential. Much family affection is exhibited, but it is difficult to judge its genuineness because southern races are so much more demonstrative than northern ones.

Dwelling houses are well adapted to the climate. The better class of houses, especially in Argentina, are built around a central court or patio. The streets of the cities are narrow, and houses are usually built close to the street so there is no space to grow trees or foliage. The city looks desolate at first, but presently, through an open door, you see palms and roses and other flowers. If you are permitted to enter, you find yourself in a sort of conservatory without a roof, or at certain seasons, in the

*Home of Colonel Bryan.*
—*H. B. Merrill photo*

*Shutter covered house at Barbados.*
—*H. B. Merrill photo*

houses of the wealthy, the space is covered with a glass roof. This open garden serves all the purposes of a living room with us. In it, the families gather to read, sew, lounge, or smoke.

The rooms are often much too high, since the distance from floor to ceiling may be the greatest dimension of the room. I suppose it is this feature that makes the rooms seem inhospitable. Windows are large, and open from the outside of the house. There are no screens to keep out flies, only shutters. In Barbados, which is as tropical as most parts of South America, the houses appear literally shingled with shutters.

In the country, houses may be small, often possessing only one room and having thatched roofs and bamboo walls. Dirt floors are common. In the cities a native wood is used, which is white and hard. Floors are washed everyday. Only a few rugs are seen, so houses look bare and unfurnished at first.

The furniture in the houses of the well-to-do is commonplace and heavy. It is usually of black walnut, often elaborately carved. One certainly sees many cabinet shops, but articles are invariably copied from European patterns. The dressing tables and stands all have marble tops and are similar to our machine made furniture. I saw one attempt at originality in an exhibit of native handicrafts at São Paulo. A number of pieces on exhibition have inlaid patterns of flowers in what I suppose is intended for natural colors; in any case the flowers are in vivid red, yellow, and blue, and the leaves in green. The effect is startling, to say the least. A writing desk shown at the same place is of a native wood, light brown in color, inlaid in white. It is good, but too elaborate. The colors are brilliant. Occasionally one sees good native carving, crude but interesting, partly because it is done in the beautiful native woods. They are much used, although they resemble black walnut, our own native wood. There is a center table of black walnut with a plush album on it. A whatnot, quite New England in pattern stands in one corner. The bric-a-brac is chiefly Majolica ware, far from English or Bavarian china, but most attractive.

# Life in the Consulates

Bedrooms in the houses of the wealthy are furnished much like our hotels. One room I was in had black walnut furniture, settee and chairs, upholstered in cerise satin. There was a cerise satin bedspread and a crimson Brussels rug. Brussels rugs are the height of elegance.

In the reception room, the settee is placed at one end of the room and the chairs are arranged in rows, or form a sort of semicircle from each arm. The settee is the seat of honor, just as at the dinner table the seat opposite the hostess is where the guest is placed.

Houses seem to be built absolutely without closets. Bedrooms are filled up with wardrobes, sometimes two or three in a room. Indeed, the wardrobe is the characteristic piece of furniture of South America. Pianos are almost as common as with us, but classical music is not heard as often as at home.

It is, perhaps, not quite fair to contrast natives' homes with those of the English and Americans, but the dining room of the English consul at Asunción is so simple, and so effective in its treatment, I cannot refrain from mentioning it. The room is large, with the bare white floor covered with seven or eight spotted jaguar skins, while on the walls are a number of old English prints. Even with this simplicity and numerous servants, housekeeping is not easy. The English lady, of course, makes tea and asks the servant for cups. The request is literally complied with, but cups only are brought in. She seems quite undisturbed and remarks; "They can never learn, and it is so hard to tell them always. I will get the saucers myself."

The minister to Brazil, Colonel William Page Bryan, recently transferred to Switzerland, and Consul General Seeger are the only ones that I see that are housed in a manner befitting the dignity of the country they represent, and their homes are much inferior to those of the German and English representatives. True, salaries are inadequate, and for this reason our

*Entrance hall.*
—*H. B. Merrill photo*

*Dining room in South American consulate.*
—H. B. Merrill photo

interests are not looked after with the care and intelligence that other foreign countries show.

Colonel Bryan's house is exceedingly attractive. He has 16 bedrooms, 12 of which are furnished so that he can entertain large house parties. The small city of Petropolis, where the legations are located, has limited hotel accommodations.

The dining room is the handsomest room in the house. The library tables are strewn with photographs of Colonel Bryan's journeys in South America, while the walls of the reception room are lined with autographed letters and other trophies that would make a college girl green with envy. One letter in particular is in the fine copper plate handwriting of Pope Leo XIII.

The office of Minister Lord at Buenos Aires presents a great contrast. It is in a remote part of the city and is so small that it will with difficulty seat three people at a time, while in the middle of the floor is a small kerosene oil stove. It certainly seems insufficient for that great country that "never has been licked and never can be licked," as they say down there, and is in pitiful contrast with the apartments of Señor Zaballos, who represented Argentina as minister at Washington during the Cleveland and McKinley administrations. Señor Zaballos has the most beautiful office furniture I have ever seen. He has traveled extensively in the United States and speaks pleasantly of Milwaukee and the Pfister hotel.

*Example of Merrill's articles sent via ship and published after the fact in installments.*

# The Dumont Coffee Cartel

The Santos Dumont's are among the wealthiest people in Brazil. They owned large coffee plantations which they sold to an English syndicate before the boom in coffee "dropped out" and hard times came on. They are modest, unassuming people, and call themselves "backwoods" Brazilians. Dumont's sister tells me that during her girlhood when they lived on their farm, or "fazenda," she was permitted to go to the city of São Paulo only twice a year. She is extremely kind to me, and in response to a request that I may be permitted to see her home, replies cordially, saying that her sister had once visited New York City. While there, she had asked to see the interior of a North American home, but because of lack of acquaintances had been obliged to content herself with hotels. She is therefore pleased to gratify the wish of a foreigner to see a South American home.

The architecture of the house is European, as are many of the modern houses, especially of São Paulo. It is one story high, built of red brick, put together with white mortar, and as if to atone for the lack of a central garden or court it has beautiful, wide verandas. The house looks much like our country houses.

The entrance hall is small with just room for a hat rack and a chair. The dining room is a dignified room as most South American dining rooms are, because the ceilings are high, and the heavy furniture suits them. The table is of black walnut and on it is a vase of scarlet blossoms. They are exceedingly well arranged, which is unusual in South America. Flowers are ordinarily arranged in a pyramid of mixed blossoms with a perforated paper disk around them, such as we used to see at the German markets. About the room stand three black walnut sideboards, a crimson brussels rug partly covers the floor, which is of a dark wood, and still life paintings of fruit are on the walls reminiscent of the Dutch masters.

The bedrooms are furnished with black walnut single beds. There are no closets, but a number of wardrobes stand about the room much the same as in France. Opening from the bedroom is a dressing room with characteristic white marble topped washstand, double in this case, and furnished with two solid silver washbowls and pitchers. The luxury of these last articles impresses me greatly at first, but I find later that silver bowls and pitchers are much used. The room for the two pretty little dark-eyed girls of seven and five is large and furnished with dramatic canopy-topped high-postered beds. The four boys, the youngest only ten years old, are in England, being educated.

The most interesting, and I think I may say beautiful parts of the house, are the kitchen and pantry. The floors are of red tile, and the sinks and wainscoting of white tile; and the black cook, a Negress, is dressed in white, and is as spotless as her kitchen. I have never seen more exquisite housekeeping, although the only servants employed are a cook-maid, laundress, gardener and coachman. The personal supervision of the mistress is evident everywhere.

*Furnishings typical of bedrooms.*

—*H. B. Merrill photo*

73

*Merrill wrote some of the same general information with additions to Birge as to the* Milwaukee Sentinel.

# The Dumont Family

During one trip to São Paulo, I was a guest of the Santos Dumont family. He was the pioneer aviator who experimented with powered dirigibles *[later the successful monoplane* Grasshopper *in 1909]*.

The Dumonts are the wealthiest coffee plantation owners in Brazil. They are a modest and unassuming people who started out as small farmers. They lived on a "fazenda" in the backwoods and went in to São Paulo only twice a year for the festivals. Now they have "palacio" built of brick with verandas and a beautiful courtyard. Every bedroom in the "palacio" has solid silver washbowls and pitchers, and the house is immaculate.

The Dumont's plantations are as vast as the wealth accumulated during the years of the coffee boom. But Señor Dumont admitted to me that he had predicted a plunge in the market and sold out to an English syndicate at a propitious time. This is one of many examples of *mestizos* profiting at the expense of foreign investors. Under any circumstances, South America, with its vast mineral deposits, timber and fertile soil, could be the most advanced country of the future.

Yours truly,

H. B. M.

*Daughters of Santos Dumont in front of their home.*
—*H. B. Merrill photo*

*Article to the* Milwaukee Sentinel

# Cultural Customs and Contrasts in South America

The North American is prepared to see strange customs in South America, but what he actually does see quite exceeds his expectations. It seems almost impossible that people who in dress and general appearance are so much like ourselves can differ so widely in little things.

The combination of luxury and lack of luxury is surprising. For instance, while you find electric lights and fans almost as common as with us, electric bells are rare. The common way of calling a servant is to clap your hands. On the boat you open your cabin door and clap your hands and a "moso," as the steward is called, appears. It gives you a sort of "Arabian Nights" feeling, and you would try rubbing the lamp to find what form its genie would take in South America, only it is an electric bulb and obviously would be of no use. In our hotels almost everywhere except at Rio de Janeiro and Buenos Aires we clap our hands for any service we need. Often it necessitates pacing the halls in our bath robes in order to make noise enough to be heard.

There is also an almost total lack of doorbells, but clapping the hands suffices. When we call on our friends, yes, even when we return the call of the English consul at Asunción, who, though we have no letter of introduction, pays us the compliment of offering his services because we speak the English tongue, we stand at the door of his charming house, built and formerly owned by an American, and clap our hands.

At Buenos Aires, at the homes of the wealthy, a servant stands at the open front door to announce guests and show them to the "sala" or reception room. This custom probably explains the lack of doorbells. It is amusing, however, to see neighboring servants out on the street engaged in the happiest kind of visiting instead of attending to their business, or standing like statues as English servants do.

It is quite customary to call women both married and unmarried, by the first name. I am not an apt pupil in learning this peculiarity, for to the end of my visit "Mees Enriqueta" (Anrekata) has to be repeated several times before I realize that I am the person addressed.

Another page of Arabian nights is opened when we discover that the men servants (most of the servants are men), sleep around in the halls and outside our bedroom doors. Our first discovery of this state of affairs is made when in Paraguay. We have dined with the American consul and his wife and returned to the hotel under their escort about 10 o'clock in the evening to find the doors locked and bolted. After some vigorous clapping and knocking we manage to gain entrance, but are covered with blushes and confusion at the sight of the hall strewn with cots and occupants. At another hotel we have occasion to rise rather early one morning and are again embarrassed to find our table waiter performing his toilet outside our door. It is difficult to figure whether we are being protected because we are without escort or simply objects of curious disposition. Actually they are the enigmas.

H. B. M.

*A frequent situation, but one Merrill did not tolerate.*

# Stopping the Streetcars

You do not have to learn either Spanish or Portuguese in order to travel on the streetcars. If you wish to stop them you stand on the corner and say "sh" as in "show." The "sh" there is quite as effective as whistling on your fingers is here. When you wish to leave the car you also say "sh". When you wish to get off the trolley, you all go "sh". When the streets are wide enough for two streetcar lines, they pass on the left instead of the right side of the street. When you pay your fare you are given a ticket which you must keep for another conductor may ask to punch it as often as he pleases. The man who wrote "Punch brothers, punch, punch with care," etc., must have lived in Argentina.

*—from H. B. Merrill collection*

Another queer custom is the sign used for renting houses or rooms in Buenos Aires. In that tax ridden city all signs are taxed according to their size. The effect on the signs is good, the neat brass plates used, are a great improvement over our glaring boards, but every effort is made to evade taxes and the "room for rent" people have succeeded so well that almost the only sign used is a piece of paper tied to the iron grating of the window. The paper may be white paper, manila or newspaper, and may or may not have writing on it. I can not learn the particular significance of the different kinds of paper.

The sign for a barber shop is a saucer with one side cut out, dating from the days when a saucer to hold the soap was fitted under the neck of the patient. It must be as ancient as our sign of the striped pole, for we read of how the erratic Don Quixote mistook a barber's saucer for a helmet.

*White paper on window grating shows room is for rent, as seen in Buenos Aires.*

*—H. B. Merrill photo*

# Selection of Food

To be sure, it is possible to get only two square meals a day, but a dozen courses are served at each meal, so that with the numerous lunches the traveler does not go hungry.

The European custom is followed by having coffee and a roll in your bedroom about 6 or 7 o'clock in the morning. That is not so bad, but when instead of coffee and rolls it is tea and toast, or, worst of all, chocolate and cake, you long for your native land. I ask for eggs, but when the bill comes in, making them worth 30 cents apiece, I desist, concluding I can survive on chocolate and cake.

Breakfast, "aimuezo", as it is called in Spanish, is served about 11 a.m. It is an elaborate meal. The queerest thing about it is the first course, which consists of a cold meat or a salad. The following menu, served at the Hotel de Paris, Posadas, a country town on the Paraná river, is a fair sample.

When I was about to sail to South America, I recall my mentor telling me, "Now Miss Colombia, go capture those crustaceans!" Only the Indians seem to include those varieties in their diet. They are not like our large crabs captured in New England and eaten with such gastronomical gusto.

The menu for dinner is mostly beef with perhaps a few more courses. Fish and soup are also served, and there are many vegetables. A favorite kind of greens is made from a tender young thistle. Dinner is eaten about 5:00 or 6:00 in the afternoon. Between breakfast and dinner, tea is drunk at 4 o'clock, and again there is tea at 8:00 in the evening. Men smoke cigarettes at breakfast and dinner, between courses, as well as after the meal. On rising from the table the phrase, "buen pro veche" is used, which is about equivalent to "now good digestion wait on appetite and health on both."

### MENÚ

*Fiambre (cold meats)*

*Cold Headcheese and Tongue*

*Sopa (soup)*

*Consommé (beef)*

*Entradas (between)*

*Puchero (boiled beef)*

*Kidney and Bacon (boiled on skewers)*

*Minuta Bifstek*

*Postres (dessert)*

*Dulce Membrillo (quince marmalade)*

*Bananas*

*Coffee*

*Note the longhorns with knife and fork at either side.*

# Two Distinctively National Dishes

"Puchero" and "mandioca" are the two most distinctively national dishes. Puchero is boiled beef, probably the meat used for soup stock. It is sent to the table with boiled vegetables, such as cabbage, beets, beans, and the national potato, "mandioca." It is like our "boiled dinner," only the beef is fresh instead of corned. The only grievance one has against it is that it is served twice a day the year-round. Mandioca tastes a little like sweet potato, but is white instead of yellow. There are several kinds, and it is eaten in a number of ways. It is also made into flour and used for bread.

Next to beef in popularity comes chicken. I am sure I've had chicken twice a day for six weeks, and at one place in the country, chicken is served course after course. First chicken with black beans, then chicken with French fried potatoes, and again chicken. Eggs are as plentiful as chicken and come in all degrees of freshness, fresh, extremely fresh, and new laid. To be sure, if you order the last, the chances are you will wait for it to be laid, but when it comes it is what it purports to be. Omelettes are good. Unsweetened they are meat. Sweetened and covered with wine they are dessert.

The cooks on the boats and in the hotel are usually either French or Italian, and food is well seasoned, a little pepper, a little garlic, plenty of oil, but delicate and well flavored, far better to my taste than most hotel cooking in our own country.

I believe the English often long for a juicy joint, but I find Italian macaroni satisfying. One kind that is especially good is made of paste *[pasta]* cut in squares, stuffed with meat and herbs and served with a tomato dressing.

The most unique dish I have had is parrot. The meat is dark, but sweet, tender, and juicy (perhaps because of their fondness for oranges?). Game birds of all kinds are plentiful.

H. B. M.

*Article to the* Milwaukee Sentinel

# Fiesta Days in Rural Areas

One must see the Paraguayans in their country setting to see them at their best or worst. A North American, Mr. Edward Peck, kindly assisted me as his country woman guest to gratify my curiosity in this respect. He invited a number of his English and American friends to a sort of house party at his fruit farm several hundred miles from Asunción. A contrast in customs begins at the railway stations. From village to village each one of them, no matter how small, has an out-of-door lunchroom. Food is peddled in abundance, both cooked and raw. A customer comes up and buys a whole roast chicken and goes off eating it, apparently considering the whole fowl a mere mouthful. Cows are standing along the platforms and beside them are tables laden with tin cups. The buyer picks up a cup and the power of the cow fills it. The milk is drunk and the cup returned to its place. Cigars are sold in large quantities. A large kind is used exclusively by the women who started to smoke during the war. The men smoke a small cigar, about the size of a cigarette. Fruit is sold, and a native candy made of glucose and nuts and quince marmalade is a favorite. Candy is poor and expensive in South America. Imported French candies and candied fruit cost from $3 to $5 a pound in gold.

Mr. Peck arranged our visit as a general holiday for his people, for a large landowner cares for his laborers in feudal fashion. The whole government of Paraguay has operated in this method to make the natives dependent upon its control. The peons come for miles around and sing and dance, and eat and drink—a respite from tedious labor. Many came rolling in on oxcarts to the ranch.

The music to which they dance has

*Carts pulled by oxen are seen everywhere.*
—*H. B. Merrill photo*

*Peons arriving for rural festivities.*
—*H. B. Merrill photo*

*Cart with natives, businessmen in street.*

—*H. B. Merrill photo*

much repetition, and it is usually written in a minor key. The instruments are guitars, violins and a harp. The players often accompany themselves by songs, which are part chant and part recitative. One of the musicians is a native school teacher who finds mathematics and geography something of a stumbling block. At least he finds it necessary to consult our American friend, a geography professor, for information as to "what that thing they call the equator is?"

The dancing is dignified and graceful, not at all like a cakewalk or a jig. Some dance by stomping and clicking castanets in a very controlled manner. The effect in the flickering candlelight with the pulsating rhythm is almost seductive. The children and babies are sleeping around the room, while the dancers occupy the middle of the dirt floor. The Santa Fe is perhaps the most characteristic dance. It is a little like the lancers, with much individual dancing, and a great deal of movement of the arms and body. The Spanish polka is also pretty. It is more like a waltz than a polka. Both men and women smoke their cigars while dancing, and the men dance, wearing their ponchos and their machetes thrust through their wide leather belts.

Mr. Peck of necessity adapts himself to native methods of housekeeping. His kitchen is a small building, separate from the house. There is no stove, but fires are burned in the middle of the floor, and by the door stands the inevitable mortar for grinding maize. These mortars differ from those used by our North American Indians, in that they are made of hardwood, and not of stone.

Usually the mortar is a log cut across and hollowed out. It is about two feet high, and is supplied with large wooden pestles. The houses in the country are small, not containing more than from one to three rooms, and usually having a dirt floor and thatched roof. There is a great lack of furniture, which is probably conducive to easier housekeeping. Hammocks are much used. One sees babies sleeping in their hammocks, swung at a height that would be a menace to the life and limbs of white babies. Hammocks are woven out-of-doors and some of those offered for sale are beautifully made. The entire event was to me a good example of the true culture of the average life of a South American family and reminiscent of our own rural areas. I was reminded of the days when we took a hay wagon to a rousing hoedown on the farm!

*Article to the* Milwaukee Sentinel

# Water, Water Everywhere—What to Drink?

  The Río Plata, estuary of the Paraná region, is second only to São Paulo in producing the greatest bulk of coffee and cerveza (beer). Throughout South America we have also been served a lot of hot chocolate and tea. How well filtered our water is, or its source, is questionable. The wine brought to the table is diluted with water. So far I have adequate liquid intake by way of the many good fruits and boiled water. On the riverboats, there are stone and brick filters which are indispensable for the Plata River systems are muddy. However elaborate any of the measures taken may look, my modern science knowledge demands more protection against the myriads of microbes that lurk everywhere in the country.

  I have longed for a glass of milk and have only seen leche dispensed directly from the cows, which are taken door-to-door in some towns, while the recipient waits for her order to be filled on the spot in whatever container is available. In the villages, butter is made by jogging the cream carried in jugs, by burrows. Without washing or salting, it is squeezed from a dirty bag and is sold for $1 a pound. Dr. Babcock would be aghast! I have had no milk on the boats. The boiled water never cools and the bottled water is charged with sufficient carbonic gas to make it unpalatable so that the

*Various bombillas and artifacts for* yerba-mate *now part of the Milwaukee Museum Ethnological Collections. Acquisitions from H. B. Merrill estate.*

traveler is ready to sell his birthright for a drink of fresh water. How I long for good old Wisconsin artesian well water.

The national beverage, or that which is used throughout most of South America, is called "yerba-mate," made from the leaves of shrub that belongs to the holly family. The tea is drunk from the cup it is made in, which is an elaborately carved gourd with silver mounts. All cups made of silver with a gourd shape are for best use. They are called "bombillas."

The dry leaves are put into the "cup" with sugar and boiling water, and when it has steeped, a silver tube with a filter on the end is used, as we use a straw, and the tea is sucked in with such solemn ceremony that one hesitates to disturb a yerba-mate drinker during his "devotions." As to the physiological effects of mate, some physicians say that it aids digestion without the bad reactions of caffeine. Europeans tell me it is very habit-forming and that a constant use suppresses the appetite. It is certainly pernicious in a number of respects. The tea is taken so hot through the hot metal straw that comes in contact with the teeth that one sees many natives with front teeth missing. The cup, or bombilla, is passed with the same tube from mouth to mouth, and undoubtedly as it cools, transmits ulcers and other contagious diseases. With the bombillas, teatime in South America is all day!

With the offering of a drink from boiling hot water, I felt it safe to try. It is similar to a drink of English or Chinese tea and is served with as much ceremony as the tea rituals in Asia. The British of course, have brought their own tea services and usual rituals, including pouring milk in the cup which makes it cloudy. I have photographed some "bombillas" and am sending some native pieces to the Milwaukee Museum to add to my artifacts from South America.

H. B. M.

# Dr. Cruz—Medical Leader Extraordinaire

Río de Janeiro

July 10, 1903

My Dear Cousin,

At last I have met someone who epitomizes all that I find intellectually stimulating as well as physically attractive in a man. To have spent a few hours in the presence of this admirable gentleman leaves me with a longing to share my life with such a mate. I met him through Professor Pedro Alfonso, a director of health here in Río. He is Dr. Oswaldo Cruz, a Brazilian—tall and slender—with dark wavy hair. His eyes were reserved in their gaze but so mesmerizing that I found it difficult to have eye contact with him without experiencing an emotional reaction. Professor Alfonso says Cruz is very shy by nature. He also said that the Dr. thought I was very young but Cruz is nine years my junior. Though shy, he moves and works with great dexterity and one feels very assured in his presence—another observance I find commendable.

Alas, having said all of this, I must add that Dr. Cruz is married to the daughter of a wealthy businessman here in Río. Knowing of his son-in-law's interest in medicine since a youth, he set up a laboratory in the basement of their home and also paid Cruz's fee to enter the Pasteur Institute in Paris. Oswaldo Cruz, the son of a country doctor, had graduated from the school of medicine in Río with a major in bacteriology. When he asked me about my work, I found him to be completely understanding of its importance to our lakes and water systems because of his own interest in microbiology and bacteriology. His doctoral thesis was on the transmission of bacteria through water. As I learned more about his work and his responses to mine were with such sincere interest, I felt it a pity that he is married. It is no surprise that he has six children.

At the Pasteur Institute, Cruz studied the same sciences that I had, but with experts such as Alexandre Yersin and Pierre Emile Roux. They proved that the diphtheria bacillus produced a toxin which led them to the production of an antitoxin. This should have been used in Milwaukee. Here I was, with a doctor who is actually involved in the histologic changes caused by life-threatening diseases that are only beginning to be taken seriously among my associates back home. Dr. Cruz wrote several monographs on everything from the plague to cholera—to yellow fever and even the toxicity of the common castor oil bean. Remember how the little tots loved to choose colored ones to string into necklaces? They often put them in their mouths. I shudder to think that those which were poisonous killed innocent children.

*Dr. Oswaldo Cruz (1872-1918) cleared Río de Janeiro and other leading Brazilian cities of yellow fever, bubonic plague and smallpox.*

Every port I have arrived in has warning flags of cholera, the plague or "Yellow Jack." It seemed to break out in Santos and spread to other cities up the coast. Professor Alfonso set up an emergency laboratory on an abandoned plantation in Manguinhos just outside of Río. He needed supplies and, in desperation, cabled the Pasteur Institute for help. "Come at any price," Alfonso told Pierre Roux. Roux answered that Río had its own specialist, Oswaldo Cruz! Cruz responded immediately and they became a team.

Dr. Cruz organized laboratories in an isolation hospital in Santos as well as the plantation at Manguinhos. He described how he made serum from killed bacilli and injected all persons exposed and ill with the antitoxins. All of his patients improved. His next effort was to warn the cities that the plague was carried by rats, and he was trying to get control of diseases through quarantine systems. I told Dr. Cruz about the many Americans hospitalized when I first docked at Pernambuco. What lives the plague didn't take, malaria did. He was not surprised that I had encountered warning flags, deterring entry to almost every major port of my destinations. A newspaper laying on the table in a cafe where I was hav-

*Horse drawn machinery resembling fire engines were used by Cruz's "mosquito brigades" to clean out all potential urban and suburban breeding sites for mosquitos and their larvae during his intensive campaign against yellow fever.*

ing coffee had a headline; "Yellow Fever, the Ruin and Curse of Our Land." Newsboys shouted, "Brazil, shunned as Land of Death." (It still continues.)

The most astounding feat Dr. Cruz accomplished was to recruit medical students who were eager for experience. He taught them how to blow their own glass ampules and make all laboratory equipment they required until the government was convinced that they needed an adequate facility for research. The students spent days and nights producing vaccines and serums at Manguinhos. To further control the plague, Cruz told all local authorities to destroy all rats, beginning in Santos. There have been close to 400 deaths from the plague since I arrived in Río.

It was difficult to imagine this shy unassuming doctor lashing out at government officials, giving orders to quarantine ships, rat-proofing and sometimes burning warehouses. If the orders were not carried out, Cruz advertised in all publications that he would pay for all rats delivered to him and destroy them himself. When I asked him about the results, he laughed, saying, "They are being delivered to the extent that I don't think I can financially or physically keep up with the offer."

The new president of Brazil, Francisco da Paulo Rodriguez Alves, is an admirer of Dr. Cruz's theories and has appointed him the Director General of Health. It is certainly advantageous to Cruz's career to have the president behind his cause, particularly since so many businessmen claim loss of profits from closing of ports and warehouses—but it is working. His present cause is to rid Río of yellow fever in the next three years. His method is to divide the city into districts, appointing a medical chief for each section who in turn recruits medical students and laborers to check all reservoirs, pools, gutters—all places where mosquitos breed. Dr. Carlos Finlay, of Cuba, and Walter Reed, a Virginian (Johns Hopkins) who has been in Cuba, are working on the *Stegomyia* species of mosquito. The

*Cruz's "mosquito brigades" consisted of paramedical personnel and laborers who worked under a staff of physicians, medical students and health inspectors. So intense was their zeal that they came into conflict with a public enraged by their tactics.*

Army Board is attempting to eradicate yellow fever so that the men working on the Panama Canal can continue it to completion. The death rate of men in the American Army Corps of Engineers and natives of Barbados is high. Both doctors are impressed with Cruz's "mosquito brigade." He organized teams of inspectors dressed in impressive, official white uniforms who are given authority to check private business places and homes, spraying them with chemicals. The object is to kill the larvae of the mosquitos and it is successful—much to the objection of the many "cariocas" (citizens) who consider it an invasion of privacy.

I have never known a man who works with such zeal to quarantine and lead a public education crusade for health, and he is working almost single-handedly to stop the suffering caused by typhus, diphtheria, tuberculosis, scarlet fever and measles. His concern for the children pleases my soul. If only we in the states had as broad a conception of a more comprehensive system of compulsory vaccinations with revaccinations every few years. As in certain cases in the USA, some citizens feel it is almost a corruption of their independent rights. Here in South America, the effort is often thwarted by the national church hierarchy. This talk of using vaccines has also become a political issue. One mad mob attacked Dr. Cruz and forced his family to flee at night into the country until military action quelled the revolt. He continues to use what the citizens consider extreme measures, and the war on yellow fever is being won.

Dr. Cruz is headed for Belém in the state of Pará which I recall as being a deadly location. I feel so fortunate not to have become discernably ill as yet and doubly fortunate to have met Dr. Cruz, which I'm certain must be obvious to you from my descriptions about him.

Tell the girls at Downer about my letter regarding these men and their triumphs. I understand I am to meet some outstanding women doctors and will write about it for the *Milwaukee Sentinel*. Watch for the article.

Love to all,

Hattie Bell

*Article to the* Milwaukee Sentinel

# The Continent of Great Potential

While South America is an open sesame to North America, and South Americans have great respect for the United States, the actions of some of our people in their country could merit contempt. The country itself has endless possibilities of developing its rich resources, and although many areas were settled here long before North America, few advancements have been made through initiative and organization by the Latin and native populace.

The soil is fertile, the climate delightful, there are splendid waterways, minerals and timber. But native South Americans seem to demonstrate a decidedly mañana mentality that is not reliable for business acumen. In the larger cities it is noticeable to a greater extent. Life is all a game of chance, lotteries flourish and one is besieged on every corner at every turn. Whole stores are given over to selling the tickets and they are sold on trains and steamboats. The lotteries are supposedly under government control but the inequity of it is excused on the premise that the proceeds go to institutions such as hospitals and orphan asylums. Ironically, the passion for gambling extends to the poorest of children. You can see them matching pennies to buy sweetmeats that are sold by numbers. Numbers are also turned on wheels that sell chances for possible prizes.

I was surprised to hear a well-to-do Latin gentleman remark to me the other day that, "It is such a pity that the Latins drove the English out in 1806." Then he added hopefully, "Uncle Sam should annex us someday!" One can't help but notice the orderliness of the attractive homes and gardens of the English which makes me think his remark was mitigated by the egalitarian principles of the British more than the USA.

In view of the recent Venezuelan question, one wonders where the Monroe Doctrine will lead us. So far, South American republics have found difficulty in governing themselves and look to us to assist in settling disputes. Germany recognizes the great wealth in the country and has a large interest in trade and immigration. They are in a competitive position with the USA. Arbitration may prevent the problem from becoming as troublesome as the Philippines or maintaining the "Open Door" in China, but it is an issue that will soon press for solution.

Just recently, the British bombarded La Guaira—particularly the fort—in response to the harsh treatment and confiscation of property owned by many British in Venezuela. The Europeans first took their cases to the Venezuelan courts and got no response so appealed to their home governments. Germany, Britain and Italy blockaded Venezuela this year. Good old Teddy Roosevelt stepped in and forced the issue to court at The Hague which forced the European national's claims to be scaled down. Bully for him!

# Expatriots Hide Specious Backgrounds

I must report that one colony at Santa Barbara, near São Paulo, Brazil, is pathetic in the extreme and in certain respects a dispensation of providence. After the War Between the States, certain expatriates from North America, who refused to take the oath of allegiance, headed for South America hoping to raise cotton as they had in the states. Many Africans had been sold from their tribes to South Americans for laborers. The colony did not prosper however, because of lack of workers. Some went to work in the cities, but the majority remained in Santa Barbara and the Italian, Turkish and Portuguese emigrants could not work with those left in the colony who refused to recognize a productive ethic.

The soil is fertile, but they do not bother learning how to maintain or use available tools. Instead, they walk the rows to be planted, digging holes in the loamy soil with their great toes, dropping seeds from pouches into the ground. It is a feckless method that nonetheless engenders enough beans and corn which constitute the mainstay of their diet.

The present generation can speak no English, cannot read or write in any language, are hardly able to sew sufficiently to keep their wretched rags together. They are a thriftless, shiftless people, with their half-thatched houses tumbling about their heads.

A charitable visitor from the southern states who has taken pity on them has established a school and is trying to have the younger generation educated; but the apathy among them is such that her work has been almost a failure. There are other colonies equally forlorn scattered throughout Brazil. To the other extreme there is a wide diversities of cultures, each asserting to establish identities in South America, the way the United States had developed during the period of expansion westward. European families, mainly the early Spanish dynasties, proudly flaunted their ancestral heritage. Most prospered through honest labor. Some hid their specious backgrounds in effecting grandiose business schemes or profitable marriage alliances in the ever hopeful quest for "El Dorado."

The Argentine Republic is today what Canada was for North Americans in the past. One's heart aches for two-thirds of the America colony at Buenos Aires and especially for the wives and daughters who have followed husbands and fathers to this new, strange country, only to learn that the name they bear no longer belongs to them, and slowly to realize that they, too, must take an assumed name.

Another instance is that of a father and son living in Buenos Aires, acknowledging their relationship and yet bearing different names. They are

*House of Expatriots.*
—*H. B. Merrill photo*

descendants of a distinguished colonial governor, but the son is known as Winthrop and the father as Thorpe.

In some cases the offense committed has been slight and, possessed of a little more moral courage, the offender would have remained at home and lived down the trouble, but these are perhaps the most to be pitied. The shifty eye and nervous manner indicate that confidence in themselves cannot be restored even in a new country under a new name.

*Article to the* Milwaukee Sentinel

# Debtors Many in Paraguay

One of the most common causes of seeking a new clime is debt, and the debtors have largely congregated in Paraguay. One dollar of our money can be exchanged for $10 there, so the American debtor leaves his debts behind him, converts his personal property into cash and lives in comfort, or even luxury, in the delightful climate of Paraguay.

The most unpardonable case is that of the villain who continues in his villainy. But the Argentine Republic has only herself to thank when her own people become the victims. A most curious bold-faced attempt at obtaining money under false pretenses is that of a man "without a country" who went to a South American bank to borrow money. He was asked for security and gave the name of a wealthy Argentine gentleman, E. B. Martini. He was told the security was acceptable and the papers were made out. In the meantime, he went to a North American man living in Buenos Aires by the name of E. B. Martin, explained the situation, and offered him $500 out of the $5,000 check for which the note called.

When the note came due, Señor Martini denied all knowledge of it. The borrower of the money was notified and was apparently much surprised that Señor Martini should have been notified; the name was plainly that of E. B. Martin, a little obscure as to the ending, but that was due to poor penmanship. Mr. Martin was produced. Yes he was sorry, but he could not pay the amount due. The bank had absolutely no redress and the man who got the money is living in Buenos Aires today.

The picture of the American colony in Buenos Aires is not all dark. There are many most charming people who are good Samaritans to their less fortunate countrymen, and the so-called American church under the care of Dr. McLaughlin, commands the respect of the community. Strangers often inquire as to the need of missions and missionaries in South America but the work Dr. McLaughlin is doing for North Americans who are in trouble calls for the support of all North Americans.

With a thrill of deep emotion one recovers self-respect and pride in country and race, on hearing in that faraway land in the English church, the prayer, "Most heartily we beseech thee with thy favor to behold and bless" not only the English king but the President of the United States of America.

*Articles to the* Milwaukee Sentinel *on Business and Industry—South American Economy*

## Factories are Few

Of course, there are few factories. Where the raw material is cheap and perishable it would seem as if they might succeed, but the difficulties are many. First, fuel is expensive; second, labor can not be depended upon. There is no danger of a concerted strike. Each man strikes when he feels like it, that is when he has a few days' food ahead. One employer told me he needed four men to perform a certain piece of work. He employed sixteen, putting them on in relays, timing them so that when he paid the last men, the first who had exhausted their money were ready to return to work. The third difficulty is that the machinery would have to be run by foreigners. The South American has little mechanical taste. In one of the large hotels in Buenos Aires a system of steam heating had been put in and widely advertised but the building was cold. On inquiry it was explained that they had not been able to find any one able to run the furnace. Again, in an agricultural school the pupils were cutting grass with machetes, and again on inquiry it turned out that they had a good steam mower from the USA but it was out of order and no one could fix it. There are some successful cotton and soap factories. I even had a nice bar of Jap Rose glycerine soap (better than Pine Tar soap) in one hotel. There are flour mills and I met the owner of one of the largest who said he put in Allis *[Allis-Chalmers, a heavy machinery company of Milwaukee, Wisconsin]* machinery a dozen years ago and it was in good condition today, even though it is difficult to get persons who are trained mechanics to repair things.

## Opportunities for Investors

Crackers are much used but are all imported. Breweries are anachronisms placed amid dense tropical foliage. They are primarily owned by Germans. The extremely hot climate doesn't seem to exhaust the industry of German arbeit! Sugar is rather extensively manufactured. It is fairly good but not well refined. The making of meat extract offers a promising field in Argentina, Paraguay and Uruguay, but on some estancias, the hide is removed and the flesh left to rot on the ground. There are beef packing establishments, but like everything under South American control, they are badly organized. The Buenos Aires newspapers printed a long story to the effect that Chicago packing houses had refused to admit Argentine meat to their packing houses, only on the grounds that Argentina was too serious a competitor. The story seems most improbable because the packing business is most profitable but it was printed with startling headlines here.

The great wealth of the country at present is agriculture. This is well-known of the coffee of Brazil, although Brazil is just now suffering from overproduction or government manipulation, at least there is a serious financial depression there in import volume.

In Argentina the crop yielding the largest returns is a kind of grass or clover called alfalfa. Wheat, corn, barley and linseed are also raised. Cattle raising is even more profitable than agriculture proper.

I was fortunate enough to be present at the great annual cattle show at Buenos Aires and I certainly never saw more beautiful horses and cattle. There are splendid horses, heavy draft animals, beautiful carriage horses, and polo ponies—for polo is much played by foreigners.

The cattle show is a gala season for Buenos Aires. There is a week of horse racing and festivities. The hotels are crowded with wealthy "estancieros" (ranchmen), who expect in disposing of their farm products to obtain the money which must carry them through the year. For one brief week even the lotteries are forgotten in the excitement of horse racing. There is a good deal of unoccupied land and it is possible for the foreigner to take up land under the following conditions. Any stranger over 22 years of age may take up one square mile of land provided he remains on it cultivating it for a term of five years. In case of death his representative may take his place. He must plant at least 100 trees, dig a well, and build a one-room house.

*Gaucho on the Argentine pampas.*

*A favorite pastime—the bullfight. Bull is being gored by the matador while toreadors stand by. In days before candid cameras there was difficulty defining action.*

# Good Land Comes High

The value of land differs in the different provinces of Argentina depending upon the quality of the soil and the distance from means of transportation. My general impression is that good land was worth from $60 to $100 an acre. The following figures were given to me by an Argentine gentleman: in the province of Buenos Aires land was worth about $35,000 in gold, a square mile. In Misiónes $10,000, and in Cordoba $5,000, a square mile. I quote the highest figures given me.

Paraguay is more than an agricultural and cattle raising country and the forests are equally valuable. Importing foreign cattle is not a success, because the hide is too thin to resist insect pests so native cattle are retained. They have great, spreading, wicked looking horns like our Texas cattle but are really quite docile. A North American dentist who has invested his surplus money in a ranch tells me that at certain seasons he can buy a cow and calf for $5 gold and in six months can sell the cow for $9. He expects to become a millionaire in five or six years, and will then return to "God's country," for most North Americans claim to have no thought of a permanent home in South America. The soil is wonderfully fertile and crops are enormous. Paraguayan tea or mate is the largest export. Grass is green all the year round. Cotton furnishes three crops a year and is planted only once in five years. A certain brown bean produces continuously throughout the year. Oranges may be purchased for one cent a dozen.

Natural resources in South America are great. Probably their extent does not begin to be understood, but industrial conditions, the fluctuation of the money market, the unreliability of labor, is such that the capitalist must hesitate before putting money into enterprises here. Investments pay enormously when successful but the amount of capital required to float them is large.

The first thing to contend with is the government, which while seeming to invite foreign capital, really does everything to hamper it. This is especially true of Brazil. Every foreign business venture is taxed to such an extent that the shrewd Anglo-Saxon organizes his company under a native name. This was done by the New York Life Insurance Company, which did a large business in Brazil, but was so taxed that it decided to discontinue business, then really reorganized as a native company backed by New York Life capital. These facts were told me by a young American business man residing at São Paulo.

There seems to be no liability insurance companies, and when I ask about it the same gentleman told me there is no need of any because no injured person could possibly obtain damages in South American courts. The laborer is completely at the mercy of his employer and his resources are so limited that it cannot be otherwise. The professions and the army are the chosen occupations of the South American. The streets are dazzling with uniforms and doctors and lawyers flourish everywhere. But the dentists, if we may consider dentistry a medical profession, are the nabobs of South America. The common fee is ten dollars an hour and the money may be demanded before the patient takes the chair. Of course, North Americans are most successful here, for nowhere has dentistry been carried to the perfection that it has in the United States. In order to protect natives, however, all foreigners wishing to practice in any of the professions must pass severe examinations in Portuguese or Spanish. No foreign diploma of any kind or from any institution is accepted. Photography is another paying business in which North Americans excel. There are numerous wealthy photographers, while the Eastman Kodak company has agents in all the large cities.

*Report to the* Milwaukee Sentinel *and the* Milwaukee Public Museum

# The "Spider" Women of South America

    Native laces are found in most of the countries of South America, but those of Paraguay are the most beautiful. This lace is always mentioned by travelers, but I think they can not have seen the most beautiful examples or they would show greater appreciation of it. In fineness of texture and delicacy and originality of pattern it compares favorably with most European laces. The Guarani or Indian name for the lace is "nanduti" (yan-doo-te), meaning spider web. About 300 years ago the Jesuits taught the natives to make this lace. I suppose the stitches are European, but the pattern is copied from the semitropical spider webs, and nanduti is the result. The spider webs themselves are beautiful. Some spiders weave in colonies and one sees great sheets of hundreds of square feet in one web. I suppose the beauty of spider web patterns for lace has occurred to many people, but these semicivilized Indians are the first to realize the dream. I have seen a tarantula that measured over ten inches across its body.

    The lace is made on a frame with a background of cloth for support, upon which the pattern is sketched. There is no braid or drawn work about it. It is all made with the needle. The thread was originally made from native plants, but European thread or silk is now generally used. It often takes years

*Spider web weave.*
    —*H. B. Merrill photo*

to make one piece of the finest quality—like the lace used by the Jesuits for altar lace. Today we do not see any of it in the churches. Instead it is made into wearing apparel and household articles, such as collars, scarfs, handkerchiefs, doilies, centerpieces and bedspreads. Its original sacerdotal character is shown in the shape still retained in many of the most beautiful collars, which have long, stolelike tabs. The lace is not much used by the natives, and one wonders where they find a market for it, for it is not sold to stores except rarely, but is peddled about and then one has to learn where the best pieces are to be found.

Argentina has an interesting lace also made by the Indians. It is a net darned in figures that remind one of Japanese and Chinese work. It is more crude than the Paraguayan lace but extraordinarily artistic. A Brazilian lace made by the negroes resembles torchon lace but is woven in colors. There are laces in Chile and other parts of South American of which I did obtain examples. These laces are not for sale in the stores and are sometimes difficult to get. I have taken several photos of the various laces and am bringing back many.

H. B. M.

*Nanduti lace.*
—*H. B. Merrill photo*

*Article to the* Milwaukee Sentinel

# Clothing Customs

When I was at the Phoenix Hotel in Buenos Aires, I was so cold that one of the housemaids on staff made me up a couple of flannel nightgowns. Florida Street in that city is the most fashionable I have seen. The jewelry store windows are dazzling. At Asunción the best shop was quite primitive. The proprietor adjusts the items you are looking at to buy on his own person, which leads to some outlandish effects. Imagine a man adjusting a woman's hat, scarf or purse on himself! In Río, the counters are covered in Brussels carpeting and when you want to look at dry goods, a book of samples is offered so that they don't have to take down the bolts from the shelves. When you finally make your choice, they let you try the material in front of a mirror and spread out patterns. For display of accessories, pasteboard cylinders are dressed with shirtwaists, skirts and belts, all covered with cotton veiling to keep the flies from messing the merchandise. Some of the models with plaster or china heads looked like brides with wedding veils. As for the men, the white flannel suit, silk hat and gloves are worn except for evening, and then any cut of coat but it must be black. Oh how a uniform is admired down here!

I should mention that the garb of the native peoples is far more interesting than that of the Spanish or English. Some wear linen or duck suits over which they toss a shawl rather than an overcoat on cool mornings. The negro women wear a chemise, heavily embroidered or with lace. Many cost $10. They wear white mostly and cotton skirts starched so stiffly that you can hear the rustle a block off. They have excellent posture from constantly carrying produce, furniture and washing on their heads and have a way of draping a shawl over one shoulder or around the waist that is becoming, whether it is a towel or coffee sacking. The Spanish still wear the lace mantilla always on religious and festive occasions, of which there seem to be both several times a month.

As for the wool favored by the natives, it is their vicuna or guanaco and alpaca. Made into ponchos, they are indispensable against the cold, damp, night temperatures. White ones cost $50. An

*Women with shawls carrying baskets on their heads.*
—*H. B. Merrill photo*

Indian woman handed one to me. It was so heavenly soft, I remarked that it was worthy of a mantle for the Blessed Virgin. She gave me a toothless smile and kissed the edge of the garment!

North American companies must learn to cater to South American tastes. The Chileans like bright colors, so the German and English send them fancy shawls with elaborate ornamentation for which the Indians pay an exorbitant price. The United States has to study the market but not take advantage of the buyers.

My North American apparel is a source of constant comment. I am asked if we always wear sailor hats (boaters). I do have my felt walking hat along, which provoked the remark that they had never seen anything like it or my thick-soled boots. I tried to take it as a dubious compliment! I will do my best to bring back a vicuna or alpaca poncho. Some of the natives are still weaving some cloth, but the Germans and English are buying up raw material and shipping the finished products back to South America, thus keeping up with the demand at a quicker pace. The pale brown ponchos are most expensive and soft as cashmere. Another popular commodity is perfume. Every dry goods store and even groceries have large stocks of cosmetics. Proprietors say that people go without food so they can buy perfume!

H. B. M.

*Coolie woman.*
—*Photo from H. B. Merrill collection.*

"I am getting some good photos of these great bronze statuesque people."

*Native woman with shawl.*
    —*Photo by H. B. Merrill.*

# The Fashionable Ciudadanos

Dear Mrs. Hare,

Roger tells me you have had concerns for my safety and comfort while traveling. It has been unpredictable but educational.

I want to assure you that your suggestions about taking extra petticoats for insulation was a good one. It is hot during the day and cold at night, so by taking some on and off, I am temperature controlled! The black one with the nice black skirt you made for me is absolutely essential for a certain time of the month, particularly as I am on the go with little time for changes.

Seamstress that you are, you would find the fabrics and clothes fashioned from them quite extravagant in appearance. Certainly compared to my ensembles. On board one of the larger boats, two girls from Argentina were wearing jacket suits, one red and the other brown, of that luxurious long-haired zibeline cloth. The sleeves were very large—skirts long and slim, and they both wore black silk and rubber girdle-style belts embroidered with steel beads and clasped with enormous steel buckles. They had matching hats of the same material and wore high-heeled tan slippers and open-work stockings. They had dark hair and eyes and beautiful complexions but more beautiful were their manners! Most other clothes are so covered in soutache braid laces, satins and jewelry (in the daytime) that it is almost in bad taste. I have seen women shopping in the larger cities looking as though they are going to a ball. Of course, the wealthy women travel by carriage everywhere and aren't apt to get muddied. Many women wear skirts that have detachable wide hems that can be removed and cleaned. Quite resourceful don't you think?

More later,

Hattie

*"The Fashionable Duo"*
  —*Hartridge rendition*

*Report to the* Milwaukee Sentinel

# Customs in Schools

The public schools in all the countries I visited were founded on our normal school system. Ex-President Sarmiento, about 20 years ago, brought a number of North American teachers to Argentina. Their wages consisted of "a horse, a house, and $2,000 in gold." They had many trials. The priests opposed their work and the story is told of one teacher who opened school day after day without any pupils; at last she asked to resign but President Sarmiento refused and told her to keep on. Later, pupils began to come and she built up one of the best schools in the country. The death recently of Miss Mary Graham, one of these North American teachers, was made the occasion of public mourning and all official and public buildings were closed.

In spite of their similarity the schools have many peculiar customs. The hour of opening school is early, about six or seven o'clock in the morning, and there is usually only one session. The pupils always rise when a stranger enters or leaves the schoolroom and the practice is so spontaneous that it is pretty. The length of the school year is about the same as ours but the vacations are shorter; but as one teacher explained, "We have so many feast days we must make our vacations shorter." I counted the feast days on the calendar and I think there were 70 during the year, which in a measure fulfills the desire of the small boy who wished that George Washington had 365 birthdays.

The amount of work required seems a great deal. So many languages are studied, a pupil often taking Spanish, French, English and German at the same time, that other subjects are overlooked. Science and mathematics are the most neglected. I only saw one laboratory in São Paulo and that was a chemical laboratory at McKenzie College at São Paulo. It was small and poorly equipped. This seems all the more strange because apparatus for lecture work is fairly good. I did find laboratories in the medical schools but in the technical schools there was little practical handwork of any kind. South Americans are unskillful mechanically with their hands and perhaps that is the reason they are now laying the foundation for manual training in the elementary schools. The work for only the first three grades of the country schools of Paraguay is given in their manual of study.

# Elementary Level Studies

Country school, manual training, first grade (for both boys and girls) simple work of local home industry, where possible, work in clay, wax and straw. For boys only: according to age, agricultural work. For girls only: needlework.

Second grade—for boys: construction of simple objects for home use, handling of carpenter tools, of farm implements, preparing the soil, transplanting, nursery, grain culture, tending cattle, bees and plants. For girls: several stitches, pieces of children's clothing, repairing, spinning, practice in handling tools and machines for the preparation of wool and cotton, dying wool and thread.

Third grade—Development of work of preceding grades—for girls: making of all kinds of

*Girls weaving straw.*
　　—*H. B. Merrill photo*

clothes, and underwear, mending, washing, and ironing, cooking and housecleaning, domestic, weaving, all dairy work.

The course sounds too elaborate, but I saw actual work in straw, pasteboard and needlework and rooms fixed up for washing and ironing, so it is in actual practice, in some of the schools at least.

There is an agricultural college at Asunción under the directorship of Señor Bertoni. The college has been established five years and has about 70 pupils. They were carrying on some interesting experiments, but were cutting grass with machetes, although they explained that they had a mower, but it was out of order.

South America has great faith in her schools and is spending a great deal of money in their development. It is hardly fair for a traveler to express an opinion concerning them or any of her institutions, judging only from the cursory glimpse.

# The Modern Woman of South America

In striking contrast to the society woman, is the professional woman. Travelers tell us that the "new woman" has not yet made her appearance in South America, but if serving on the boards of public institutions, organizing clubs, and taking an interest in national reforms constitute the "new woman," then she may certainly be found. Whatever name we give them, women are actively engaged in the professions. There are women doctors in Brazil and in Buenos Aires I think there are four. The one I learned to know best was Dr. Cecilia Grierson. She speaks English well and it was a great pleasure to talk with her. She has a large practice. I had occasion to visit her office several times during office hours and it was always filled with waiting patients. In addition to her practice, she is director of a training school for nurses and a member of the board of the woman's hospital. She is interested in a home for working girls, which seems to be managed on much the same plan as the Young Women's Christian Association here, and she is advocating a bill, which is pending, modifying the divorce laws or rather giving the right to divorce.

In Argentina after a woman is married, the husband is given the most absolute control of her property, person, and children, and divorce is impossible, no matter what the abuses practiced may be. I suppose the laws are needed, but so far as observation goes, home life seems happy. Much family affection is displayed and husbands and wives seem to have reached a very satisfactory understanding as to their mutual obligations. I remember one fond papa who closed his eulogy on his small son with the exclamation, "Maria, he is fine! He is in heaven."

Nor is the woman's club wanting. The International Council of Women which claims to have established branches in eleven civilized countries, of which Mrs. May Wright Sewell is president, has a branch in Argentina. I will quote a part of the report at the first annual meeting by Dr. Ernestina A. Lopes:

> *"In compliance with the agreeable duty assumed by each member of this subcommittee, that of the press of Argentina, I wish to make you acquainted with the latest events of feminine interest which have occurred in our country. The minister of public instruction conceived the idea of forming a pedagogic congress which should meet annually in February with the purpose of bringing together the directors or representatives of all the normal and national colleges in the country."*

I will omit the account of the meeting and give only the closing paragraphs:

> *"I am pleased to note that the feminine portion of the congress occupied a prominent and brilliant part in all the discussions, notwithstanding that the representatives of the opposite sex were for the most part lawyers and professors of high degree."*

> *"The Columna del Hogar," our ladies' review, started the idea of giving a 5 o'clock tea in honor of the ladies, etc. At the conclusion, the directress of the Normal School in San Juan proposed to form a society of all the intellectual women present. Later at a meeting held at the house of the president of the national council Mrs. Alvina V. P. de Sala, (five days later), the bylaws of this society were explained to help those ladies who wish to study."*

I am sure the above extract sounds quite North American. The club mentioned in the last paragraph, gave a reception at which I was the guest of honor before leaving Buenos Aires. Among the members are doctors, dentists, authors, artists, editors and, of course, teachers. The dentist is a pretty, vivacious young woman, charmingly dressed in a pale blue wool gown. Señora Catalina Allen de Bourel, the editor of the woman's newspaper "The Column of the Home," is the moving spirit of this organization. She is a large, fine looking woman, speaks English and possesses much judgment and poise. Her paper serves to keep Argentinean women in touch with the work of women all over the world. A recent column contains a portrait of Mrs. May Wright Sewell and several views of her home in "India."

Among the guests were a number of North America women, the most distinguished being Mrs. Robinson Wright and her secretary, Miss Hartman. Mrs. Wright has written a history of Mexico and of Brazil and is in Argentina getting material for another book.

During the afternoon I overheard a little conversation that I cannot help repeating. Mrs. Wright asked one of the ladies about some information her husband had promised to obtain for her and the lady replied, "I have not seen him. You know I am so busy I do not see him often." I think this shows that some South American women are quite as "advanced" as their sisters in the North.

## The Foreign Women Have Clubs

The foreign women have their own clubs. The English women of Buenos Aires have an organization for social purposes only. At São Paulo the women have formed a club to which only women who by birth or descent are North Americans are eligible. The club pin is a circle enameled in our own national colors, red, white, and blue, and in the colors of Brazil, green and yellow. Needless to say it is more patriotic than beautiful. In the same city I was fortunate enough to be present at a delightful reading club, for these faraway women keep remarkably well informed as to current English literature.

It was a great surprise, as well as pleasure, to see the pin of the Daughters of the American Revolution. They think there is no chapter in South America but they hope it will be only a short time when there will be the requisite number of members to form one. I wish I had more time to spend with these inspiring women but I have truly found the women of all of the cities and rural areas as exhilarating as their countries. And now, alas I will be leaving soon.

*Hattie Bell Merrill bids a renuente adiós to South America.
—scene of the beach at Ipanema and Leblon (Praia do Arpador) by M. L. Harrridge*

# Part III

# Back on Terra Familiar

*Bascom Hill 1907. Upper left—Science Hall behind new State Historical Society building; lower right—lower campus and Red Gym. Lake Mendota is shown upper right.*
—*Photo, SHSW WHi(X3)2319, courtesy of State Historical Society of Wisconsin.*

September 1902

My Dear Hattie,

At last, you are safely home, but of late, I have noted a bit of frustration in your sense of importance in pursuing the unseen and even in your ability to find time to write of your many experiences. You claim you don't "put things as succinctly on paper as you think them." I have heard nothing but good remarks regarding your articles in the *Milwaukee Sentinel*. My unprofessional guess is that it is not your research or writing that has confounded your confidence, but someone or some influence acquired in South American that altered your viewpoint.

I have been reading Henry James again (from the *Art of Fiction*), I think it is apropos to your endeavors, particularly the use of the word "unseen."

> *"The power to guess the unseen from the seen, to trace the implications of things, to judge the whole piece by the pattern, the condition of feeling life in general so completely that you are well on your way to knowing any particular corner of it—this cluster of gifts may almost be said to constitute experience. If experience consists of impressions, it may be said that impressions are experiences." Therefore, I say—write from experience only. As James said, "I should feel this rather a tantalizing monition and would carefully add—Try to be one of those people on whom nothing is lost!"*

Your concerned but proud, Roger

*Ledger cards from Board of Regents Executive Committee Report Vol.8, page 216, February 28 and April 10, 1910. Request regarding Merrill's position and salary as assistant professor of Zoology—Hauks moved—seconded by President G. D. Jones. Granted $600.00 for part-time position to work up results of South American expedition.*

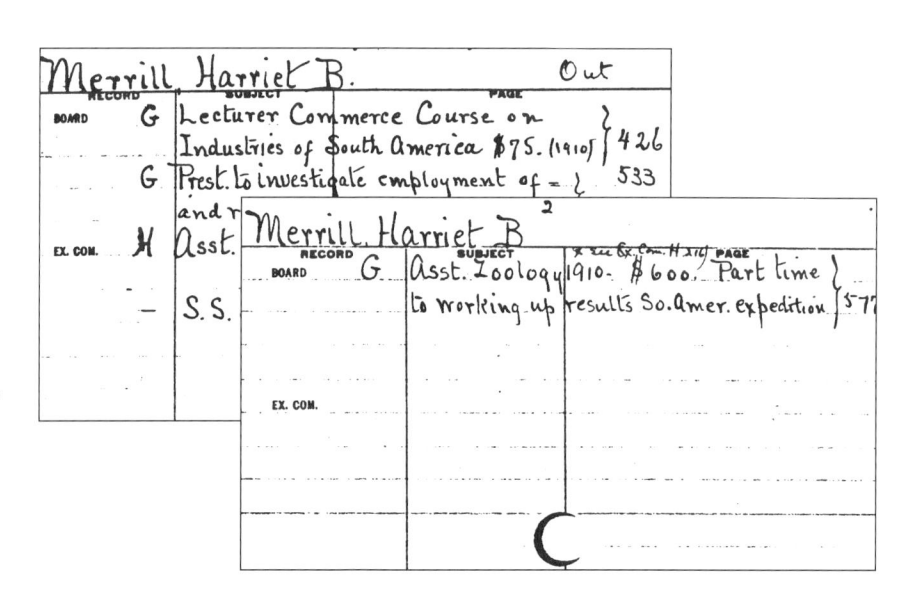

*Refer to chapter notes 3, page 193: Academic Criteria—Early 1900s*

# Harriet Merrill's Return to Appointed Tasks

*As a member of the faculty, Merrill was aware of the internal political factions that existed alongside the administrative processes at the university. When President C. K. Adams suffered a breakdown that forced him to leave Madison for the South and Europe, Birge had been appointed acting president. At the death of Adams, Merrill felt it was logical that Birge should be advanced to the presidency. She was aware of the opinions expressed by his adversaries on campus, who claimed his policies would be a deterrent to progress in promoting the Wisconsin Idea. (Refer to chapter notes 6, page 196: University Politics)*

*Prior to the election, Merrill launched a petition advocating Birge as the man for the position. As vice president of the Wisconsin Academy of Sciences, Arts and Letters and the Wisconsin Alumni Association, she wrote to alumni groups urging their support. In a letter to a Milwaukee alumna, Miss Northrup—November 17, 1902—Merrill stated that she had conferred with Birge on the matter of the presidency, although he was concerned about the feelings of alumni opposition.*

November 17, 1902

My Dear Miss Northrup:

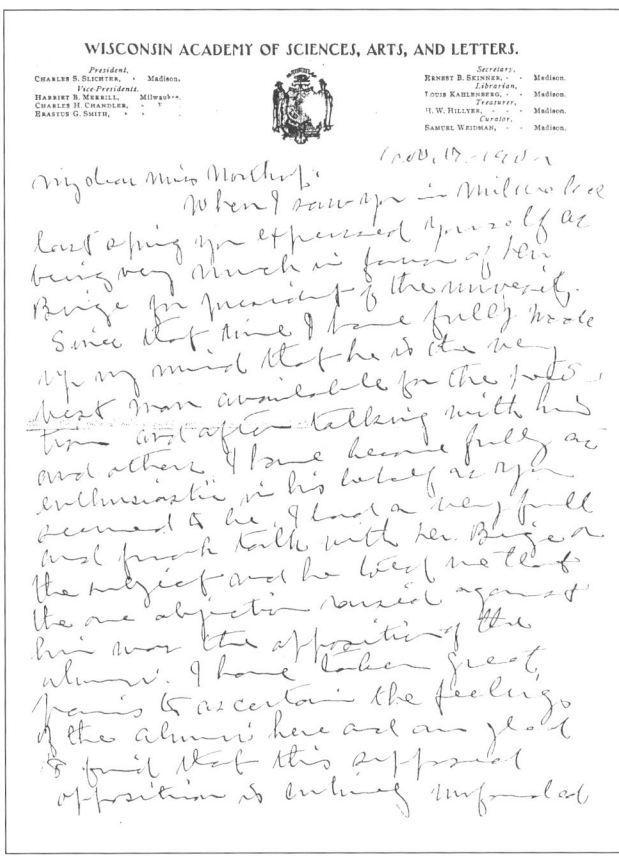

When I saw you in Milwaukee early last spring, you expressed yourself as being very much in favor of Dr. Birge for president of the university. Since that time, I have fully made up my mind that he is the best man available for the position, and, after talking with him and others, I have become fully as enthusiastic in his behalf as you seemed to be. I had a full and frank talk with Dr. Birge on the subject, and he told me that the one objection raised against him was the opposition of the alumni. I have taken great pains to ascertain the feelings of the alumni here and am glad to find that the supposed opposition is certainly unfounded. We are now preparing a petition addressed to the committee of the Board of Regents for the selection of the president, urging the appointment of Dr. Birge. I thought something of the kind could also be done in Milwaukee. I have, from very high sources, the assurance that such petitions will unquestionably bring about the selection of Dr. Birge.

It is quite important however, that this be done as quietly as possible and that nothing regarding the matter get into the papers. I thought I would write you to see if anything could be done with the Milwaukee alumni.

For your information, Vilas, Bascom…? *[the writing becomes difficult to decipher]* are all enthusiastic supporters.

Very truly yours,

H. B. M.

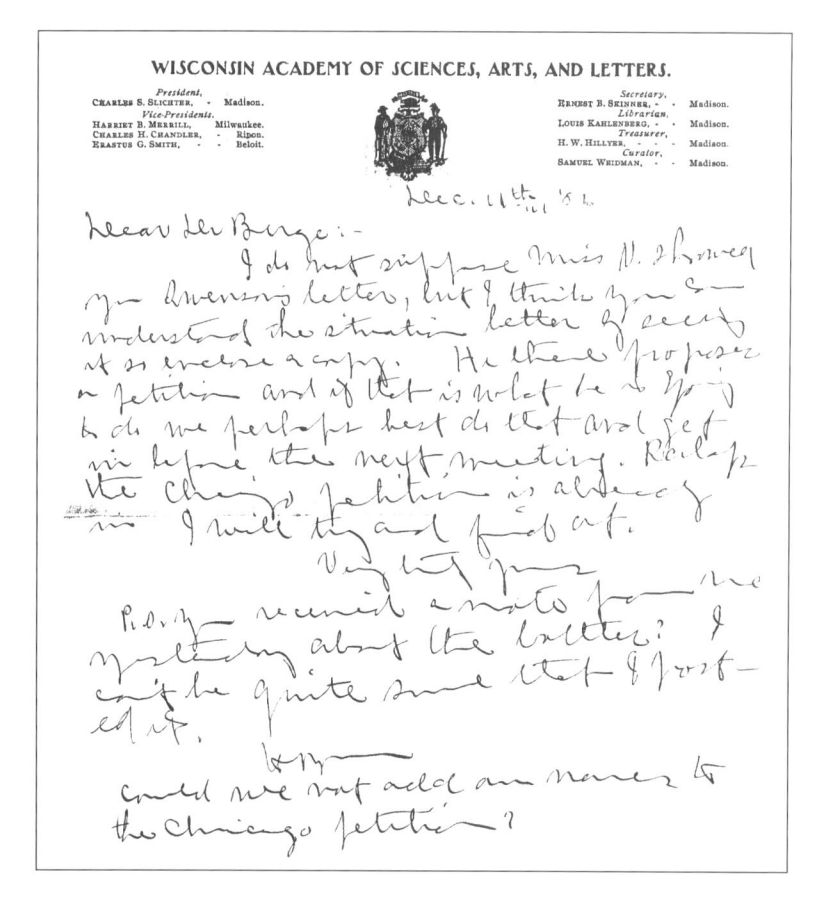

December 11, 1902

Dear Dr. Birge:

I do suppose Mrs. N. [Northrup] showed you Swenson's letter, but I think you can understand the situation better by seeing it. Enclosed is a copy. He therefore proposes a petition, and if that is what he is going to do, we perhaps best do the same and get it in before the next meeting. Perhaps the Chicago petition is already in. I will find out.

H. B. M.

P.S. Could we not add our names to the Chicago petition?

# 1904–A Historic Year for the Capitol of Wisconsin

On February 27, 1904, fire-fighting companies including Milwaukee's No. 1 with a team of three white horses, hastened to Madison to extinguish fires in the third Wisconsin State Capitol. Some of Merrill's students assisted brigades of volunteers in relays to retrieve valuable records from the blazing building.

Woman in white shawl, Mrs. Louis Hess, rushes from the scene visible from avenue off the square, to her home. Note the hook and ladder teams standing-by at the curb. The photo was taken by a newspaper photographer.

The fire was started by a gas lamp that had been installed so high up in the cloakroom, that it eventually caught the ceiling on fire. Firemen were called but they could not get water from the capitol hydrants and by the time they were able to get water from city mains the fire spread to the south wing and dome. The fire ruined 80 percent of the capitol.

—*photos from M. L. Hartridge collection*

*Students attending Milwaukee Public Museum, Milwaukee, Wisconsin.*

—From H. B. Merrill collection

## The Intriguing Legacy From Young America

*Between lectures at the University of Wisconsin, Universities of Chicago and Cornell, Merrill lectured at the Milwaukee Public Museum. This was a letter typed for a lecture at the Milwaukee Museum and later sent to brother Roger for family history.*

Dear Roger,

I should think your interest in the early history of the New England colonies would be stimulated by your being in Boston so often. You may recall that I lecture regularly at the Milwaukee Public Museum and the curator Henry Ward asked me if I would give a talk on my D. A. R. background. Surprising to me, the audience was attentive and interested. They particularly took note when I mentioned that Brunswick, Maine, was heavily forested and populated by the Algonquin Pequot tribe and that our great grandfather cleared land and was appointed to survey the area which was East Andover, Massachusetts, before it became part of Maine in 1820. Our own grandfather Roger became a trusted friend of a Pequot brave called Nattalock. He helped the family locate the best trapping and fishing places.

Reading over our genealogy in preparation for my talk, I was again reminded that great grandfather, a lumber operator, insisted that though Brunswick was situated in the dense environs of a lumber and seafaring town, his sons should have an education the equivalent of Harvard. Many of the Merrills were graduates of the U. S. Military Academy at West Point, but almost every Merrill family had a student at Bowdoin. When Bowdoin College was founded in 1794 all eligible Merrill sons attended. I pointed out to the audience that there were 11 children born to Sarah Freeland, daughter of Dr. James Freeland of Sutton, Massachusetts, and four of our uncles attended school wearing handwoven flax shirts, woolen pants and jackets with bone buttons, and leather boots. Everything except the metal buckles on their shoes came from their land. I showed the coin silver spoons and some china that great grandmother had brought with her from her home in Sutton, Massachusetts.

Squire Merrill's wife was described as small in physical stature but strong in courage. There were few neighbors before the college brought people into Brunswick and the house that Roger constructed from his own timber, was built for his large family, and could accommodate friends for stopovers. It became almost a wayside inn.

This reminded me of Henry Wadsworth Longfellow and Nathaniel Hawthorne, who according to the family were at a symposium at Bowdoin, and visited the Merrill house as a guest of Uncle Leonard Parker Merrill who was a Bowdoin graduate in 1842. In reviewing the lives of the early New England settlers, I found a similarity in our paternal lineage to many of the men of Bowdoin. Family notes describe

*Booklet from The Public Museum showing Merrill as lecturer.*

—From H. B. Merrill collection

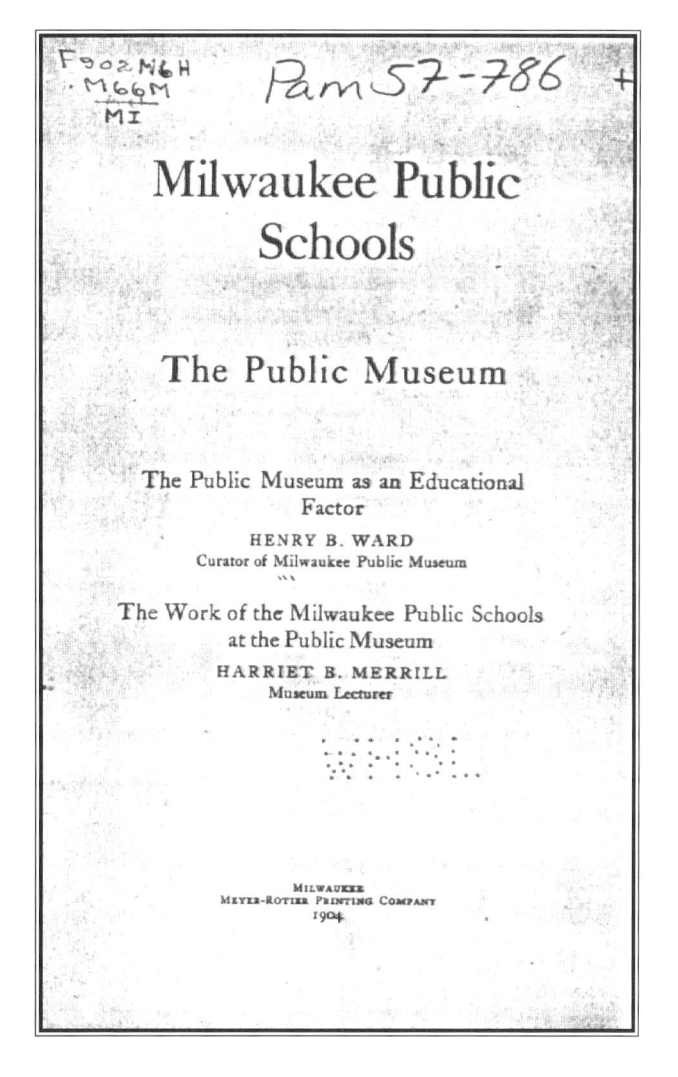

Longfellow's interest in learning that the Merrills often opened their doors to the Indians. The house had become sort of a refuge for wayfarers from all paths and since grandfather logged in Pequot country, a need for trust and assistance had to exist on both sides. I told a couple of stories you may not remember that are on record. There was a main large room in the middle of the house with an enormous fireplace. When the weather was unbearably cold, the Indians were sometimes allowed to spend the night in the hall or the big room where they preferred to sleep stretched out on the floor, heads pointed toward the fire in tribal custom. In return they brought the squire's wife some supplies but usually she made the trek to get them on her own. There was only one general store in the area where villagers could buy household goods—a trading post at Bethel, near East Andover, Maine, eight miles down the path. In good weather this was a short run with a wagon. But one spring day Sarah decided to walk to the store and was stranded in a sudden unseasonable blizzard. Rather than wait it out and with concern of the children back home, she borrowed snowshoes and through an almost obliterated trail, she managed to bring back provisions.

Just when I thought I had talked long enough the audience wanted more. I told them about grandmother's trust in being alone when the men were off at the mill or at school.Sarah was often alone with the youngest children and felt threatened only from wolves or bears but one day a marauding band of Indians tried to break into the house. They had never come with such boisterous actions. They had been drinking spirits and were totally out of control. Leaving the children inside she went out to the back kitchen, separate from the house where she had been baking bread, and yelled at the Indians to go to the rear yard. She put a pistol in her apron pocket and waving a loaf of bread in one hand and the pistol in the other, she told them to sit down and she would boil up a pot of chicory which they would have to drink before getting food. Perhaps she hoped it would leach out the rum they had consumed and sober them up!

Henry Ward allowed that I could display the excellent firearms collection that had been in the Rudolf Nunnemacher acquisitions at the museum. There were some fine pieces admired by the men in the audience and the guns added a graphic dimension to the subject, particularly a powder horn and flintlock gun just like grandfather's. I told the class about Great-grandfather Ezekiel Merrill, who was assigned to report at Lexington, Massachusetts, as a minuteman in 1775 and fought in the first battle on April 19. They lost nine minutemen that day. He later joined forces with colonists who beat back Major General John Burgoyne and his Canadian recruits in 1777. The audience was all at attention.

When great-grandfather died, his widow was left to fend for herself and she managed her land as long as her strength allowed. As a testimonial to her the proprietors of Andover voted her a section of land in her own name. She selected Number 1 Y Range which she sold to a brother for $100. With the money she bought ten large Bibles, giving one to the church library. This dissertation may not instill in you the feeling of belonging to the past as it does to me. I often wonder what it might have been like to have stayed in Brunswick in which the aura of two of the most gifted writers of the century, Longfellow and Hawthorne, commingled with Poe, Lowell, Irving, Emerson and others—all learned men, many of whom interacted with one another and had connections at

*Bowdoin crest.*

*Leonard Parker Merrill, an uncle of Harriet Bell Merrill.*

*—Photo from H. B. Merrill collection*

Bowdoin. Not all had traveled the world and were as keen in linguistics as Longfellow but all had a gift of interpreting certain aspects of America that we may never see again except through their literary comprehension of this region and its heritage.

Nathaniel Hawthorne's father, Nathaniel, was a sea captain out of Salem and his grandfather an owner of ships before the revolution. Longfellow's background touched the sea at Portland.

Both men were of the class of 1825 at Bowdoin although they were not close friends until later in life. They certainly must have been an influence on Uncle Leonard Parker Merrill. He received his M.S. degree from Bowdoin in 1842, was a member of the Bowdoin Athenaen honor society and an honorary member of Delta Kappa Epsilon. Hawthorne was also an honorary member of Delta Kappa Epsilon and the Peucinian Society. Longfellow was a Phi Beta Kappa upon accepting a chair at Bowdoin in Literature and Linguistics with instructions that he must travel abroad and learn as many languages as possible. I was surprised to learn that his full beard was the result of attempting to cover severe burns to his face sustained when he tried to rescue his wife whose dress ignited from burning sealing wax. He rolled her in a rug but she died. It was the day of their wedding anniversary that she was buried!

I am looking at a photograph of Uncle Captain Leonard Parker Merrill. He studied law with a judge (Bowdoin had no law school) and practiced for a time while married to "the lovely

*The Merrill home, Brunswick, Maine.*
*—from H. B. Merrill collection*

Caroline Kent Newman," and they had two sons. There was an overabundance of lawyers in Maine during this period while activity in shipping had accelerated after the War of 1812. Many British seamen who claimed American allegiance had been captured at sea and our own shipbuilding companies were in need of commanders. Leonard watched the tall conifered cloisters on his father's land cut down for masts until it was almost stripped bare. Some men in the audience expressed concern over the excessive cutting of Wisconsin forests.

Shipbuilding was in full production, and although it was said it was a vexing situation for Leonard to leave his settled existence, he took command of a fleet of merchant ships with plans to see the world and return home in a few months. Probably inspired by Herman Melville and other adventurous men whose lives bordered on the sea, he sailed off around Cape Horn and having contacted yellow fever, he put into the port of New Orleans where he died in 1870. His body was returned to Brunswick.

Looking at his photograph one can discern a firm set of the chin, the stern straight forward look in his gaze and feel that he was capable of commanding the men and his ships. The mishap of the last crossing was sad. I treasure his picture given to me by Aunt Bella *[Arabella Merrill]*. He also handed down the Merrill Heraldic banner which had a cross from a crusade on one quarter of the shield.

I wish that if you go up to Brunswick you would look up the old Merrill house. I am told it is still there. The furthest east I will get for awhile is Urbana, Illinois.

I miss you.

Love,

Hattie

P.S. I can't recall the number of children in the family of grandmother Emmons but know that father was one of eleven born to Roger and Sarah Merrill. We haven't carried on the tradition have we?

*[Merrill family genealogy filed in the New England Hannah Goddard Chapter of the D.A.R. The New England Merrills, published December 9, 1987, Brunswick, Maine. A Merrill Memorial—Samuel Merrill: An Account of the Descendants of Nathaniel Merrill–Early Settlers of Newbury, Massachusetts, published 1917, Cambridge, Massachusetts.]*

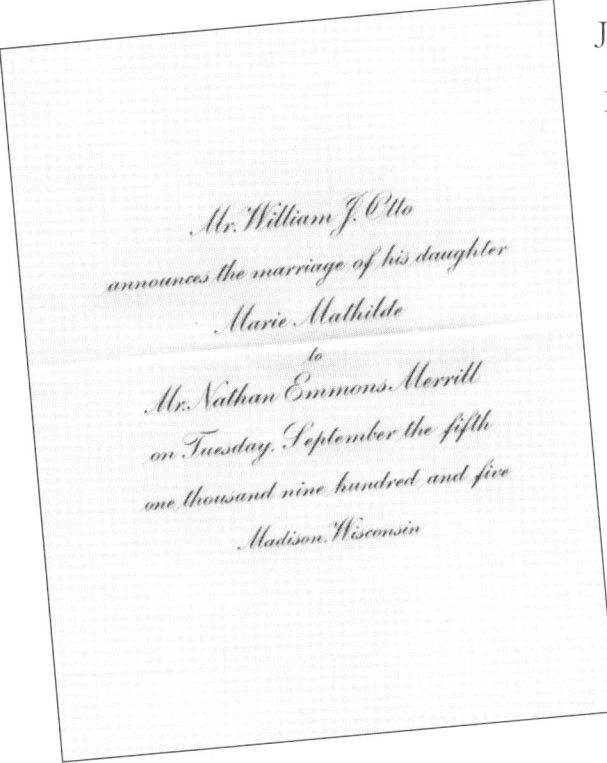

June 1910

My Dear Nate,

While I was out of the country, you and your betrothed were married and oh, how I wanted to be there! When I first met Marie (her family calls her Mary?) I was taken by her efficient, poised manner. In the wedding picture she looked tall, slender and quite regal. There is a lot of gray matter (I don't mean hair) beneath her auburn pompadour! My wish for you Nate is that homemaking will be as fulfilling as her success in the business world has been. From what you have said, she seems willing to remain in Chicago for as long as it is important to

*Wedding invitation of Nathan and Marie.*
*—from H. B. Merrill collection*

*Marie and Hattie.*

—From H. B. Merrill collection

your work and her's. Do you think she would mind giving me some advice on investments?

I have been invited to visit Marie's home and also with her married sister's family this coming week. I will write to Roger about it. I must post him a line of thanks for taking me about in Milwaukee when I was there last. I met him at his office in the imposing new Majestic Building (next to the Merrill Building). His company has as much space there as in the home offices in Baltimore, he said. Busy as he is, he took me to dinner at the Pfister in what is referred to as the Fern Room.

Do you recall my study of the species of *Filicinae* and all the photographs I took of ferns for the Chicago St. Paul Railroad Co. publication? It seems now that certain factions made up of naturalists have to be convinced that the rail systems have no intention of eradicating all growth in the paths of the roads. Incidentally, the paucity of ferns in the Pfister Hotel should cause no protest. If they intend to call it the Fern Room, I would suggest importing a boatload from Brazil! In any case, it is a fine hotel and convenient to the Milwaukee Club right across the street. I am so glad that Roger has comfortable quarters, with all necessary amenities, in such a reputable private club. It amazed me as to how many persons I met in South America were familiar with the Pfister Hotel.

I am pleased that you like the silver picture frame. One can always use an extra table stand frame. It was intended for you and Marie when I saw it in a shop (Regalos de Maria) on the Calle Victoria, when I "Rolled down to Río," sin hermanos—solo!

Again, I hope for you both the very best and was truly disappointed to miss a major event in your lives.

Love to you and keep some for Marie,

Hattie Bell

# All American-Germans

June, 1910

My Dear Roger,

I have finally had the opportunity to pay a visit to Nate's father-in-law, Wilhelm J. Otto. Marie's sister, Johanne (Hannah), met me at the streetcar stop and we walked to the house situated just off a lovely little park surrounded by old Victorian-type houses. Hannah's grandfather Johann Otto came to Madison in 1850. He was a founder and charter member of St. John's Lutheran Church and was "ein leiter" in this active Germany community. Family members served in the Civil War.

I didn't know that Hannah is a nurse. She has been taking care of her father whose diabetic condition has deteriorated since the death of his wife, Wilhelmine, five years ago. Last year the youngest of the siblings, and only son, died at age 27. Johann—named after his grandfather—had taken a severe cold and Hannah, knowing he had a fever, told him not to go bird hunting in the damp marshy areas. I gathered that he was an impetuous youth like so many who think they are invincible. He refused to listen, got worse and died of pneumonia. Hannah said that her father took the death of his son harder than the loss of Julia, her twin sister, when they were 18 years old.

When her mother was expecting her birth, she was thought to be carrying a boy, due to her large size. The family chose the name Johann for "him" and when the mother delivered twin girls, one was named Julia and she was named Johanne. Julia died of consumption, which Hannah says is too prevalent in this robust dairy farming state. She is well aware that there is still a lack of regulation in controlling the distribution of raw milk and thinks that was the cause. Dr. Babcock would agree.

I expected to find a somber atmosphere at the Otto home, but when we arrived, pleasant strains of piano music were flowing from the bay window of the front parlor. Hannah's younger sister Adlaide uses the room as a studio occasionally. We walked down a long dark hall that has an elaborate hat rack with bench, and through heavy double doors into a comfortable, carpeted sitting room which was surprisingly quiet. Another pair of oak sliding doors muffled the sounds emanating from the studio on the other side but I don't believe there is a door made that could stop pipe tobacco smoke from permeating a house!

Mr. Otto got up from his overstuffed platform rocker where he had probably been dozing. With newspaper and unlit pipe in one hand, he managed to make the usual polite salutations. He did go on somewhat about Nate's work as a patent attorney, saying that the company he was associated with has been developing new methods for increasing productivity. He then added soberly that he would probably never see the advances accrued as a result. He is a semi-retired machinist engineer for an Electrical Manufacturing firm in Madison and attempts to manage two family properties near the lake, one right on Lake Monona, for one of his widowed sisters and her daughter. When Adlaide heard that I had arrived, she dismissed her student and opened the doors to the front parlor which resem-

—*photos from M. L. Hartridge collection*

*Marie Merrill
(Mrs. Nathan
Emmons Merrill)*

*Grandfather Johann Otto*

*He arrived in the United States in 1848 during the Heidelberg student rebellion. Photo is a fine example of a daguerreotype with embossed velvet lining and gold leaf on frame.*

*Johanne Otto*

*Adelaide Otto*

*Conradine Hess*

bled a small salon. She introduced herself and apologetically bustled about picking up stacks of etudes off the settees and tables, and placed an exquisite paisley shawl over the large, ornate Steinway grand piano. There was an attractive selection of silk and linen upholstery fabrics which Hannah had been in the process of applying to the chairs etc. with gimp tape. Addie introduced herself, excusing what she referred to as "a mess in Hannah's attempt to remove the old black horsehair from a love seat. It would last another hundred years," she said and then pulled up two chairs to a marble-topped tea table. I had the feeling that Hannah was the family maid as well as nurse. Addie is Marie's youngest sister and a charter member of the Wisconsin School of Music, located in the university area where she teaches. Nate said he found her to be refreshingly candid and I agree. In just the short visit we had while Hannah busied herself in the kitchen, Addie filled me in on the origin of the fine etchings and other objects d'art that hung on the walls and filled the surfaces of what was otherwise a very plain and modest room. Some things had been handed down over the decades. Others (she confided in a tone almost penitent) were gifts that had been given to Hannah as payment for delivering babies, nursing families, sitting up all night with sick persons and helping the injured!

Hannah hurried into the room with a tea tray and cakes. It was obvious she had heard the conversation, probably for the 'nth time. I tried to change the subject, remarking that the tea service was one of the finest I had seen as indeed it was. Then Hannah, with an apologetic laugh, said, "perhaps the nursing profession doesn't always put food on the table but it put the tea set on the table!" A prominent judge's wife had given it to her in gratitude for Hannah's help when the elderly woman had broken her arm. I think she said the name was Burrows. To lighten the occasion, Hannah proceeded to read our tea leaves. When our cups were nearly empty of tea, she swirled the liquid around and while examining the residue of tiny brown shapes, told our fortunes which was nothing more than amusing entertainment, particularly when she "read" that there had been a man/in my life!?

Addie walked me back to the streetcar stop, apologizing for criticizing her sister and assuring me that she admires how devoted she has been in caring for the family. Hannah is the one who keeps up the property, does the shopping, takes their father to GAR meetings near the Square unless his friends pick him up with horse and buggy. Autos are scarce. He has a veteran's pension of some kind and the properties he inherited. But it is evident that his daughter, warm-hearted Hannah, is the strength in his household now. She is aging before her time. I saw a photograph of her and her twin among the family pictures. They were real beauties. There was also a picture of Johann—a healthy, curly-haired good-looking young man who bore a striking resemblance to a daguerreotype of his namesake and grandfather, old Johann.

Tomorrow I am invited to tea at the home of Marie's sister Conradine Henriette Hess. The car stops at Hess's Corners which is in a new area beyond the Yahara River. Will write you a letter about it as soon as I see them for when I get to my lectures, my time will be limited.

Love to you,

Your Hattie

# A Truly Gemütlich Gathering

June 1910

Dear Roger,

Whatever would you do without all the reading material I provide for you to read on the trains to Baltimore? This is to fill you in on my visits with Nate's in-laws.

I took the street car from the University, way out to Hess's Corners and was surprised at the number of houses being constructed since the transit system put tracks East of the Yahara River. The car went past new churches, a large factory, and I heard that Lowell School is proposed for the Fair Oaks subdivision, almost at the end of the car line. I can see how Mr. Hess speculated that with the burgeoning population, this new location should be lucrative for business. Marie says that he owns several lots and three houses uptown near the Capitol Square as well as acres of land around Fair Oaks and Lake Monona where he has three good sized summer cottages.

Louis Hess first built a temporary store on one section of his property as a trial venture. He also had two houses constructed on adjacent land. When it proved to be a thriving community that demanded more merchandise, he built a larger building with the latest well-appointed equipment and with living accommodations above. This store with a soda fountain and ice cream parlor was written up recently in the *Wisconsin State Journal*. The Hess's moved from their long-time home

*Mrs. Louis Hess.*
—*from M. L. Hartridge collection*

*Louis Hess Store at Hess's Corners.*
—*from M. L. Hartridge collection*

uptown to the store, although as Marie said, Conradine had hoped to live in one of his houses. Mr. Hess made the final decision to live above the store.

When I arrived at Hess's Corners, a couple of nicely dressed businessmen got out of the car with briefcases and walked down Elmside Blvd. toward a new subdivision on the lake. The rest of the passengers joined the farm and cottage commuters who, while waiting at the "station," were stocking up on staples. Some had brought eggs and dairy products to sell or barter. I had to step around horses, carriages and farm wagons that were tethered near the store. Mr. Hess has his own vegetable garden adjacent to the store, and it was there that I met Conradine (Dena). With a dish towel, she was shooing some ravenous robins who were attacking the cherry trees. Then, waving the towel in my direction, she rushed over and greeted me with as much decorum as any council general's wife I had met.

Like her sisters, Dena is fair complected without benefit of powder. The dark curls on her forehead were damp, and in spite of exertion, her afternoon frock looked immaculate. As I followed her into the store, I detected a subtle fragrant combination of lavender and spice which reminded me of Mother. Instead of taking me upstairs by way of the private entrance, Dena said her husband wanted to meet me before the other guests arrived at 4 p.m.

"It is his busiest time of the day," she said, "If Teddy Roosevelt walked into the store at that hour and Louis was with a customer, President Roosevelt would have to wait," she laughed.

At first glance, Mr. Hess reminded me of Dean Birge. It was the hair and the mustache. But Louis Hess is a tall, kindly, avuncular man who, though self-effacing, is known for his philanthropy. His wife Dena confided that one reason her husband has been reluctant to be away from his business for long is because of a great monetary loss he experienced through an investment in what he thought was a worthwhile cause, entered into on faith. A couple of area doctors had solicited donations from well established Madison businessmen to support the building of a hospital. It was a project that Louis felt was sorely needed, and he, among others, contributed thousands of dollars toward its completion. Located at 413 South Baldwin St., the hospital was well built, and though small, it was difficult to staff with medical personnel and it continued to lose money. Finally, the city refused to support it and wouldn't even grant tax relief. The investors, of course, lost their

*Mr. Louis Hess*
—*from M. L. Hartridge collection*

*Florence Hess in car.*
—*from M. L. Hartridge collection*

hard earned savings. The Hesses referred to it as the "Boyd case." "Since then, Louis rarely leaves the store. Sometimes he goes fishing with friends, and he owns a fine touring car but we never tour," Dena sighed. In the early evening, just before I left, Mr. Hess took some of the guests over to the building (which had been the temporary store) that now houses the splendid automobile with its shiny black upholstery and large acetylene lamps—all up on blocks!

About four o'clock, the store was bustling. Every time the doors opened and closed, bells jingled and the sounds of the coffee grinder, meat slicer and cash register must have been a welcome din to a man trying to recoup financial losses. But to the family living above, it must be an intrusion in privacy and tranquility. I learned that in this upstairs-downstairs existence codes of communication were improvised. Messages within the building were deciphered from a certain number of taps on the radiator pipes. The telephone was used mainly for business or emergencies and there was a rule that Dena was never to allow it to appear that she was tending the store alone. She was there part of every day and had to know the inventory of all groceries, school books, stationery, tobacco and other sundries. On the way to the back stairs leading to the second floor, we went through the soda fountain parlor, which was accessible from a side entrance off the boulevard. Two prettily starched and beribboned little girls pranced in and up to "Mama Hess." Of all Marie's sisters, Dena is the only one with children. They had been to a birthday party with chums in their old neighborhood uptown. She said that they were not happy with the move to "the end of the world," and I sensed that the transition was not easy for mamma either. Marie's oldest niece, Florence, is tall for a 14-year-old. She has blonde wavy hair, and when I admired her big cornflower blue eyes, she gave a quick curtsy, excusing herself to go and finish a drawing, a talent for which she seemed to have a natural bent. Her 10-year-old sister, dark-haired Elsie, is quite shy and acknowledged my presence with a sweet, dimpled smile. One of the interesting studies in genetics is the dominant and recessive characteristics passed on to offspring. Dena Hess is fair with blue eyes, but very dark hair. Louis is dark eyed with black hair. Both are from light-haired German parents. The two sisters are a contrast in coloring and demeanor. But, Louis is intent upon setting aside enough funds to send both on to higher education and from all indications expects them to succeed in their own chosen field of interest.

When we got upstairs, it was obvious from the chatter that other guests were arriving from the private entrance, and there was a tray with calling cards on the hall table. I recalled seeing this custom in South America also. When a caller arrived in person, he would bend one corner of the card. If delivered by someone else, no bent edge! I was simply amazed at the size of the rooms, every bit as large as in a standard house. Tea was steeping in a pot on the range and, rather than a tea table, a sumptuous buffet had been set in the main dining room. I have noted that when German women give a tea, it is more like a Vienna kaffee klatsche, but this resembled a dinner. Of course, not all afternoon affairs are given in such proximity to a confectionery and grocery!

*From left: Florence Lucille Hess and Elsie Elizabeth Hess, about 1909.*
—*from M. L. Hartridge collection*

Out behind the good-sized kitchen and pantry, a long hall led to a porch which spanned the width of the building. The men had taken it over, feeling freer in the openness to smoke a cigar or pipe. Dena introduced Dr. Richard Mössner to me, a most amiable gentleman. He is a Northwestern Medical School graduate who interned at Cook County Hospital and practices in Chicago. Nate knows of him. And the distinguished Arno Bierbach and his good-looking son Norton (you have met both) who are well known Milwaukee bridge engineers. They asked dozens of questions about engineers they knew of who were involved in similar work in South America. I told Norton that I was scheduled to give a lecture that might cover the subject, and I would let him know when I was to be in Milwaukee. John Krause was there with his daughter Leonore who was just graduated from the UW. He studied at the University in Heidelberg…is a violinist and plays an Antonio Stradivari, which he says he was fortunate to obtain from a musician in Cremona. After many years, he is still experimenting with components in the instrument that are key to its unique tone! (He was written about in a W. S. J. article as knowledgeable about violins and the lacquer used.) John's father-in-law was the Honorable Francis

*Hattie Bell, Marie Merrill, Florence Hess and Elsie Hess.*

—*from M. L. Hartridge collection*

Massing who was a student in the law offices of Judge Levi Vilas. He also established the first Männechor in Madison and was editor of a German publication. John mentioned that Louis used to enjoy the German Sangerfests here and in Milwaukee and wishes he would take time off to attend more.

As a Yankee, I have always been proud of my New England heritage, but I have to admit that the Germans I have seen, in both North and South America, are the most industrious lot when it comes to business acumen. Louis Hess's mother Elizabeth Mössner came from Gondelsheim, Germany in 1848 with seven brothers and sisters—married Heinrich Hess from Baden-Baden—of an equal number of siblings. Louis is the seventh of eight living brothers. The Mössners came here with backgrounds in law and medicine. Several studied in Heidelberg and came to the states during the Heidelberg student rebellion era. Others came with skills that warranted business establishments very early after arrival. They own bakeries, a cooperage, cigar factories and grocery stores.

I had some notions at times that the lives of women of German heritage are circumscribed around "kinder", "kochen" "und kirche". Where Marie's families are concerned, I couldn't have been more wrong. Dena introduced me to one of Louis's cousins who managed further introductions while things in the kitchen called for attention. The attractive and confident young woman took possession of my arm, leading me from one to another of the women who monopolized the parlor. The lady, Olivia Monona (stage name) is a recent grad of the University of Wisconsin in science and languages. The latter has helped her career, beginning at the Chicago Lyric Opera Co. and will continue at the Metropolitan, which has recognized her proficiency in speaking six languages! *[She later became financial advisor to the Metropolitan Opera and to Director Rudolf Bing.]* All of the women were UW graduates, teaching mostly in the arts, literature or languages and of course, unmarried—except Wilhelm Otto's sister Minnie, who is Norton's mother, and Julia, who is Director of the Milwaukee Women's Exchange. The whole female assemblage, right down to the starched shirtwaists and pince-nez spectacles, affected an atmosphere of intimidating, sober reserve. It is no wonder the little Hess girls approached the prim, austere gathering with shy hesitancy. When Dena announced, "essen," the men came in and Dr. Mössner enlivened the scene with his jovial manner when he asked Wilhelm Otto's sister, "How many women were exchanged in Milwaukee this week Julia?" The ladies' fans fluttered at his remark to much laughter, and the aromas from the kitchen put everyone in a good frame—as Dena directed, "sitzen sie bitte."

Although it was not a sit-down dinner, it was not a tea as New Englanders think of it either. With ample rye bread sandwiches, jellied beef and horseradish, sausages and cheeses of all kinds, pickled beets, creamed cucumbers "mit schnittlauch" and Dena's hot buttermilk "krum kuchen", it was a balancing act! Then to top it off, she insisted that everyone have a slice of her kirsche (cherry) torte just out of the oven with a dollop (her measurement) of the best vanilla ice cream I have ever tasted. While we supped—nearly surfeited—young Florence went to the piano, softly rendering several etudes taught to her by her Aunt Addie. All in all, it was one of the most pleasant afternoons I can remember. It was also an edifying group, as they all were well versed in world affairs. One cousin, William C. Sieker, who received his B.S. degree at the University of Wisconsin in 1899, had a fairly good concept of my research endeavors. He collects butterflies as Professor Freeman does, but lepidopterists cannot be expected to "fathom" Cladocera. I simply thoroughly enjoyed visiting with persons with diversified professions. It was most edifying. *[Attorney Sieker was a Mössner descendent.]*

Dena and Louis Hess were warm gracious hosts in arranging such a grand family gathering. One of the cousins had a horse and buggy tethered by the boulevard, ready to take me uptown, but I decided to take the streetcar. Mr. Hess made a kind gesture to me upon my leaving. He placed a little paper bag in my hand as we parted, saying I was welcome back anytime. I felt like a celebrity as the party at Hess's Corners saw me off. On the way back to the campus, I opened the bag and found it filled with chocolate bonbons, pink wintergreen and white peppermint lozenges and a new, violet-flavored product, a foil wrapped, roll of hard candies with holes in the middle *[the first Life Saver]*! I found the peppermints very agreeable after an evening of much "essen".

By now you must feel you have spent the day with me and are ready to close your eyes.

With Love,

Hattie

Chicago

November 24, 1911

My Dear Hattie,

When I look at the little silver spoon and fork and the beautiful napkin ring engraved, "Aunt Hattie," I can't stop crying. Our precious baby, Rawson (named for Grandfather Merrill), died yesterday after a short but noble struggle for life. He was perfectly formed, I will never forget his dear curly-topped head and fine nose and forehead. And his eyes—deep blue—seemed to gaze at me from heaven. Like our first son, he was a "blue baby." We prayed this time would be a successful birth. Something must be done about this problem with too frequent deaths of full term infants.

Nate is taking our loss and my depression as well as can be expected. He would have been as dedicated a father as he has been a husband. My nieces, whom you met at my sister's, have considered Nate the kindest of men.

We thank you again for your concern and will always think of you as Aunt Hattie. Keep yourself warm and well.

In deepest sorrow and love,

Marie and Nate

*Photo by M. L. Hartridge of objects from Merrill's adventurous life displayed on a silk and wool paisley shawl from Marie Merrill. Of special interest are the ancient Chinese inkwell, an early microscope of Dr. T. L. Hartridge and large coin silver spoons engraved with the name Emmons.*

—*Hartridge collections*

# Part IV

# The Birge—Merrill Correspondence

*From the period of May 22, 1914 to April 10, 1915 there were over 62 letters written to Birge. Based on statements in the letters, it is obvious that Birge responded promptly to all of them. Some are typed, when she had access to a typewriter. Most are handwritten and in pencil, on various types of paper ranging from full page size, some with letterhead, to strips about 4 x 9 inches in size. Far different from her letters from South America, they talk about her health problems, but also show concern with university matters at Wisconsin and at Illinois, where she enrolled for the Ph.D. program beginning in September 1914. Birge knew that she was ailing, and her doctor recommended against her leaving Oconomowoc, but she had the determination to pursue her degree in spite of any difficulties.*

*In May of 1914, Merrill decided to go into Milwaukee for a diagnosis of her shortness of breath which she had experienced since her return from South America. During the period in which she was confined at Columbia Hospital, under the care of Dr. Henry Vining Ogden, her letters to Birge and his prompt responses were constant. He saw to it that she had a considerable amount of reading material, some that he knew appealed to her literary tastes and many university journals and papers. Merrill was almost apologetic about her physical condition which fluctuated in severity and prevented her from getting about as she had in the past.*

*Merrill's letters to Dean Birge were written with the brevity owed to her sedulous academic schedule coupled with an amazing struggle to maintain her expected physical capabilities. The letters are typical of busy persons who have corresponded to the extent that details concerning mutual interests are understood as related, in an abbreviated style—they certainly showed plucky veracity!*

∞

182 Biddle Street

May 12, 1914

Dear Dr. Birge,

I have been to several places by electric car but can't stand much exertion as my heart has been a little out of gear. I will try to get into the country somewhere where the cost of living is cheaper, possibly Beaver Dam, Oconomowoc or Greenwood. Before I had to leave to come to Milwaukee, I requested that my research material on Cladocera be deposited, with all of my papers, in the laboratory at the Biology Building. Included in the papers are notes with corresponding numbers on bottled samples that were collected in British Guinea. I don't want them to get misplaced.

Yours sincerely,

H. B. M.

May 20, 1914

Dear Dr. Birge:

My physicians told me that I must find a restful location and get off my feet. This necessitated making arrangements to stay outside of Milwaukee in a home-care situation, and I am staying with Mrs. N. L. Crummey at RFD 28, which is four miles from the Oconomowoc railway station and two miles from the electric line on Dousman Road. The home was an old stagecoach inn on the route from Milwaukee to Madison, which has been a little modified since its period of use. I have a corner room with south and west windows, which makes the room light and cheerful. The food is good, and I hope to "pick up." The cost is $6 a week, which I consider quite reasonable. There are two toddlers in the house who disturb my work during the day, but I manage some writing at night until midnight when the landlady helps me get ready for bed. Your letters arrive from the station by way of the "family team." I appreciate the information on what is going on with the faculty and to know that you are storing the material that I catalogued and sent to you. Thank you for all of the books. *Pepys* is one of my favorites, also the *Atlantic Monthly*. The sad circumstance is that I have no one immediately near to share my interests. It astounds me that of my attending physicians, Dr. Ogden is the only one who is familiar with *Pepys' Diary*. Dr. Haase has never heard of him!

Yours truly,

H. B. M.

∞

June 4, 1914

Dear Dr. Birge:

I received your note before I left Milwaukee, but could not get a reply to you so sending it now. Hoping it will be awaiting you in Madison.

I think I shall be very comfortable here. The house is an old stagecoach inn on the road from Madison to Milwaukee and has been left but little changed. The rooms are numbered just as they were in stagecoach days.

Hope you had a good time at Missouri. I am not doing any going at all—keep very quiet most of the time.

Very truly yours,

H. B. M.

R.F.D. 28

June 12, 1914

Dear Dr. Birge:

 The map and books came only the day after I wrote. Thank you for both. I read John Galsworthy's first. His stories always seem a bit morbid, but vivid. I have read only a few of the R.M.—very good—certainly no spinster flavor to them. Miss Mason brought the *Scribner's* magazine and *Outlook* with Roosevelt articles—not the slightest literary touch to them.
 I will send the books back before your family returns. They *[Birge's wife]* struck good weather for Maine. Quite cool today.
 I want to contact Dr. Evans again. I want to know if my heart is any worse. The beat is more regular, but the slightest exercise sets it going. Dr. Evans has prescribed strychnine, as had Dr. Sheldon previously. I am concerned about tuberculosis. Several of the family at Oconomowoc have died of this disease but I'm quite certain none of them had lived in my room.

 H. B. M.

∞

*On June 18 she was examined both Dr. Ogden and Dr. Evans. Evans urged her to stay but Ogden said she should go back to Oconomowoc for a couple of days. Columbia Hospital did not have a bed available at the moment and St. Mary's Hospital, which did, was too far away for Ogden. All this conflict disturbed her so much that she was ill all night on her return and was unable to retain any breakfast the next morning. She managed a glass of eggnog at 10:00 and felt quite herself again.*

Columbia Hospital

June 23, 1914

Dear Dr. Birge:

 You must be provoked that I did not remain in Milwaukee as I was supposed to. I guess I didn't realize that Dr. Evans was really interested in treating me further or to supervise my condition. I thought he considered the case turned over to Dr. Ogden and was just kidding as always. I was told he was not feeling well. I will have to write to him and it is difficult as I feel that I treated the situation badly. Do you know his address?

 Yours truly,

 H. B. M.

Columbia Hospital

June 26, 1914

Dear Dr. Birge:

    I feel much better and have not blundered badly after all. I told Dr. Ogden that I did not understand that Dr. Evans was supposed to treat me and he kindly said he wanted to come and see me and wanted me to feel satisfied. He went back to Madison Sunday night I'm told. Dr. E. quoted you to me saying he lied to Dr. Ogden as an excuse to see me—referred to you as the "old goat" as usual. I wondered what you might have said to him. I told Roger to have Dr. E. meet me at Dr. O.'s office so all is clearly explained. Dr. O. is not feeling well himself! Today was hot and he was tired. He scoffed at Dr. E.'s suggestion of liver enlargement, perhaps a fluke from my travels in foreign countries.
    Your letter came this morning, but *Pepys' Diary* has not yet arrived. I haven't written to anyone else yet.

    Yours truly,

    H. B. M.

∞

Columbia Hospital
930 Sycamore St.
Milwaukee

July 3, 1914

Dear Dr. Birge,

    Dr. Haase has eliminated digitalis from my medication which Ogden felt was essential. Dr. Evans has made a diagnosis which Dr. Ogden disagreed with and the final opinion was myocarditis and pleurisy. I have no fever or cough, and though they also suspect tubercular pleurisy, the bacillus has not been isolated.
    There is to be a homecoming in Juneau, Wisconsin, on July 1-5, and Roger has been counting on me attending.
    Among the materials you sent, I saw the name G. E. Birge in a list of voters. Judge Barber of Chicago alluded to a Charles Birge and C. C. Dietz spoke of "a new house this Birge had built." I never heard the name in Juneau. He was before our time. Is he Stanley's father or kin?

    Yours truly,

    H. B. Merrill

July 10, 1914

Dear Dr. Birge:

Yes, I would like to see you again. Regent Hammond has offered to take me out and Kate McIntosh will be coming by. Of course, I can't go very far. Why can I not meet you at some interurban station before or after Bayfield? I think I can get an old, old horse they have, but one good enough to go from lake to lake I know of—go west—do not cross interurban bridge—I am about sure to get some horse. I think you will find me looking better than I have in a long time.

H. B. M.

P.S. Isn't that picture in the funny part of June *Harper's*, page 161—I think—like the "(Menden?) folks?" Quite their expressions.

*In spite of doctor's orders that she should curtail activities, Merrill managed to keep up with her assignments. On July 31 she traveled to Milwaukee to check her South American boxes, which without telling her, a Mr. Brown had shipped to Roger. She had written Brown two letters about the matter but had no reply. Roger was in Baltimore when her things arrived and Merrill was not aware of the shipment until the freight office sent her a notice that she was being charged 30 cents per day for storage. "Brown has just written about my stuff in the museum. So I shall probably come to Madison soon. It is all boxed and can be left for sometime."*

*Her health problems continued, although at a reduced level. When she traveled to Milwaukee on August 1 to see Dr. Ogden, she noted, "the little walking I did brought back the fluid so I must be quiet. It is this fluid, which Ogden considers tubercular in origin, that I have to be careful about. I cannot tell myself until it gets high enough to affect the heart, but any exercise seems to bring it on...Ogden does not want me to eat milk and eggs, nor do any stuffing."*

August 6, 1914

Dear Dr. Birge:

I will have to come in to Milwaukee to look over my things, many of which are probably missing. I am anxious to know. Roger intended to put the things in the attic of the M. *[Milwaukee]* Club if the boxes are not too big to fit there. I have to sort out several items. I have made up my mind that I want to study with Professor Zeleny and plan to go to Urbana next year for graduate study. Dr. Ogden has objected and I will let him do so.

H. B. Merrill

*Continuing her studies was a major effort for her because of her health problems. She wanted to fulfill her lifelong dream of earning a Ph.D. Even with their close working relationship, Birge was not necessarily instrumental in placing Merrill in Illinois simply because of his close association with H. B. Ward, a major instructor in freshwater biology. Ward had just moved to Illinois from Nebraska, and when learning of Merrill's health problems, he might have preferred for her to remain in Oconomowoc. Birge was well acquainted with Merrill's work in his field but took no candidates at Wisconsin. Even Chancey Juday, his research assistant, went to the University of Indiana for his degree after several years under Birge.*

*Still continuing her interest in Cladocera, Merrill asked if Birge could get her some live specimens from the Madison lakes or even from Green Lake. She had hoped to get some material herself from the lakes around Oconomowoc.*

Friday, August 21, 1914

Dear Dr. Birge:

I'm studying "those Cladocera papers," and I am glad that you are getting your papers completed and published. I will travel to Madison tomorrow to pack up my things Monday, and Tuesday also, if necessary, for shipment to Urbana. I hope to catch the train from Oconomowoc which will get me to Madison at 11:00. Failing that, I would catch a later train from Dousman, which will get me there at 12:50. I plan to eat lunch and then come straight to the University to your office or to the Biology Building. If there is any other news I will tell you when I see you.

H. B. Merrill

*The Science Department staff at the University of Wisconsin in 1912. Photo was taken at the dedication of the Biology Building located south of Bascom Hall. Woman at right is thought to be Merrill. After Birge's death, it was named Birge Hall.*

—*photo courtesy of John E. Dallman, curator of University of Wisconsin–Zoology Museum.*

*Merrill traveled in to Milwaukee and back to Madison rather frequently as her research materials were being shipped to Milwaukee for storage and she presumed the rest of the items would be shipped to Illinois. She was full of enthusiasm, remained active, and very capable in Cladoceran work while keeping up with numbers of university affairs between her several academic connections.*


1010 1/2 W. California St. Urbana, Ill.

September 22, 1914

Dear Dr. Birge:

If only I felt well I would enjoy this. The work is rather a load.
I am head of the laboratory division and teach two quiz sections in which the students, 33 each, were from household science and premedicine mixed. A horrible combination. I have never had such poor classes in understanding general announcements. The first laboratory work was to be on a nearby pond, and hence in quiz sections, I thought I would touch on your thermocline work. Ward is such an organizer. The first morning when the class went out for field work, the pencil sharpener was out of order and while doing experiments on parasites he blew up the jet main! In the laboratory he introduced us all to the class, made us rise and bow to a large group in the well-filled lecture room. Shelford who is to take over general direction is unreliable and unless I miss my guess, he and Ward will clash. I find he has made mistakes in your calculations every time I look at his book—see pages 58a and 59 from your work.

 H. B. Merrill

∞


610 W. Mathews Street, Urbana

Wednesday, September 23, 1914

Dear Dr. Birge:

I will write just a line to let you know that none of my mail was lost in the shuffle of moving. Even the one directed to Nat. Hist. building came to 610 Mathews. I was provoked at first, for I thought Ward had ordered it away, but I had other mail and the postman put all together. The room in Nat. Hist. Bldg. is 310 Room, not Box. My name is not in the directory so letters are just as well sent to the house. The two previous letters came in the same mail. I moved Monday afternoon and things are pretty bad but I am near and food is plenty.

The freshmen are as verdant as elsewhere. One instructor reported a quiz question "discuss metabolism" and got "cell theory" from one and "mitosis" from another. I got nothing quite equal to that, but I have not attempted discussions. They get "What is a cell?" "What are the physiological properties of ppm?" etc. from me.

High schools generally excuse pupils from exams and the effect is very bad. Discipline here is, as usual, carried too far if not erratic. Pupils know they will not be excused and so stay out anyway, and since they must make up, the makeup hour is a great nuisance.

I feel a little better. My cough is better and I am better when I do not strain my heart. My hours are scattered. I have laboratory 10 to 12 Tuesday, Thursday and Saturday.

H. B. Merrill

∞

September 26, 1914

Dear Dr. Birge:

Carl Haessler was here Sunday—enlightening Illinois on Socialism as it is at Oxford. His name is not in the catalogue—don't know if he has a place here or is a guest. Not much interest in his subject. Professor Moore gives Sunday talks such as yours but his are socialistic in tone. If you can get a study copy of the "unpopular magazine" without much trouble, please send one on. Goldschmidt of Munich has been here—Ward has guests and lawsuits and so far given lectures on parasites and has Kalenhey's reputation with the freshman.

Illinois beat Minnesota on October 31 which made the town go crazy and on the same day Wisconsin and Chicago played to a scoreless *[0-0]* tie. This also made the Illinois people happy because from faculty on down, this institution hates C.U. *[Chicago University]* No one tries to conceal it.

*[Illinois beat Chicago on November 14, and beat Wisconsin on November 21, at Camp Randall; Illinois 24-Wisconsin 9].*

I've heard stories by way of student conversation that belied the reputation that Urbana was a prim town. Some girls had gone to a "frat" ball, and in the wee hours the chaperon for the house was seen taking a sorority girl off the veranda where she was sick from too much champagne. Men and women were wandering about late in the kitchen, etc. The better the discipline, the better the student, the more reliable the work I always say.

Yours sincerely,

H. B. Merrill

1010 1/2 W. California St.,

Urbana, Ill.

September 27, 1914

Dear Dr. Birge:

*The Leopard* came. Thank you. President James was a Methodist at N.W. and a Methodist he remains. Never in all my experience have I had so many directions as to what to do and what not to do as I have received in one week here.

Ward has exhorted us about half a dozen set regulations and last night James addressed the faculty and students on the growth and aims of the university which this term had an enrollment of 4,400; the largest in history. Teachers <u>must stand</u> when conducting recitations. All the buildings are locked on <u>Sunday</u> and the only excuse for going to them is to feed the animals. I shall proceed to grow animals. Smoking on campus is forbidden, no golf on Sunday. Clark made a speech which amounted to forbidding dancing by the faculty. I have no burning desire to break any of their rules, and you may get an overlong Sunday letter, since I can hardly remember when I did not go to the lab on Sunday. The Catholics and Jews do not seem to be represented among the churches and the Episcopalians have a service in Agricultural Hall. Can't be very high church. I do not know whether there is a town church.

Now for gossip. Did you know that James probably has cancer of the stomach and that his wife is practically insane and in a sanitarium or asylum? They tell of queer things she did at faculty receptions and say she does not recognize her own family—very sad.

Kingsley was very pleasant, asked if I was at Wisconsin when he was, and talked a little about the war. I understand Germany is in danger of civil war. He said the Germans took big chances and had entirely misjudged the situation. He thought the emperor was forced into the position. I told him about the Vilas travails and he said when he got his ticket there were 12 first-class tickets to choose from and that they, the V's, did not know how to proceed.

I shall probably get a seminar in neurology under Kingsley. I talked with him and he spoke of you. He is a fine man, but looks quite old, like one who has been stout and then lost weight.

Boarding house is the same old story. I sit at the table with Mr. and Mrs. Paul, American Literature, German and English teachers and librarian, Miss Simpson.

It is cold and rainy today.

Very truly yours,

H. B. Merrill

1010 ½ W. California Street

September 29, 1914

Dear Dr. Birge:

Glad to get your letter of September 27 which reached me when I came to lunch Monday at 12. Pretty quick time. I am glad your centrifuging work is so promising. I have not said anything about it and will not.

I need some algae for growing more Cladocera. What algae did Woltereck use? Can you send me some? Are the first and second cards in a catalogue I sent him correct? Will you correct and return the paper? I found some Cladocera today and can't think of the name—I found lots of it in South America. Enclosed is a drawing and description of *Scapholeberis* which I have done from memory for lack of reference material. It has a long spine at the posterior end of the shell-ventral, straight indentation between head and body, antennules small and well covered in †. If I get any Cladocera I think are worthwhile I might have them sent alive.

Yours,

H. B. M.

∞

September 30, 1914

Dear Dr. Birge:

I received your letter at Oconomowoc. I know it disturbed you that I was making fun of my condition and that I shouldn't take it so lightly. By the way, I did not "poke fun" at being turned over to Providence, but I was so decidedly lame, halt, and blind that it was a little funny. I did, however, appreciate the spirit in which it was sent.

Yours,

H. B. M.

Sunday

Dear Dr. Birge:

Have been packing all day—landlady says I must pay a month in advance for lease. The girls say I may leave it to the Dean. Didn't know I had come to be that auspicious. Last Sunday I cleaned microscopes in the lab, which I have done since I first came, and put labels on drawers. Heard whistling and commotion as at U.W. but saw no one—then one boy came in. I am to have an outside key if I get up enough energy to go for it. The freshmen meeting was called partly because the University has just defended a lawsuit this fall, brought about by a proud papa, to reinstate "sonny," dismissed for poor work. The University won, I believe. Tuesday is "graduation day."
You will not hear from me again until I get well settled.

Yours truly,

H. B. M.

∞

October 5, 1914

Dear Dr. Birge:

The key came and I will return it when you like. Ward has been away for a week. I will ask about keys when he returns. Zeleny has no special problems but wants me to get and rear all sorts of Cladocera. He has some species he sealed up five years ago that are still alive and suggested working on them.
When you have time, I would like my work books (not the lantern slides) that I sorted out and another Birge net. If you have time to get any Cladocera, it would be worthwhile to have them sent alive.
U.W. got two people from here: Mills in music (said to have a very testy disposition). Does he take Louis Coerne's place? Ricker in zoology took his master's degree here last year.
Mrs. Trelease called on me again and also asked me over on Sunday. Then Wm. Jr. had to go to the hospital for an emergency appendectomy. He is alright. It was amusing to hear Mrs. T. praise Illinois. She said the University got more money from its mill tax in Chicago alone than Wisconsin gets from the whole state. Is that true? Botany is just finishing a $20,000 conservatory. Are you really worried over the Allen investigation?
Just scratching this to let you know that the key came but a good deal torn open. I rather need my books and a net. Shelford and the boys would fish for me if I had it.

H. B. M.

October 7, 1914

Dear Professor Birge:

I have had very many enjoyable times on campus at the faculty gatherings where I have met with Mmes. Larson, Craighorn and a Miss Carn, who has charge of the Episcopal House. She called on me, and we exchanged news on Milwaukee connections. I am glad you have not forgotten me entirely. In regard to the Allen Survey[8] *[the efficiency report being done at the University of Wisconsin]*, Ward spoke of the Wisconsin investigation headed by Charles Allen. Zeleny said he saw Pearse, and I could imagine him snooping around and saying we could get back at Allen now since he had published a preliminary report. Allen has looked in on classes and sent Normal people to check. They reported on the number of hours that the labs were not in use. Is this in the papers? Which ones? Do not write if you are tired or ill. I hope your alimentary canal has straightened about.

Very truly yours,

H. B. Merrill

∞

October 10, 1914

Dear Dr. Birge,

I have kept track of many young students through their families. Many men may ultimately be forced to delay their plans to complete their educations. I met the son of some Milwaukee people who is a graduate of St. John's Military Academy *[St. John's Military Academy was founded, post Civil War, by the Episcopal Reverend Sidney Smythe and three women, Katinke Mauer, Minna Fryer and Mary Schuchardt, all sisters of Dr. T. L. Hartridge's grandfather Louis Schuchardt.]* and who was to have spent a year of study in Europe. The world situation has changed since June with the assassination of the Austrian Crown Prince in Sarajevo. Studies in many cities of Europe may be canceled. The papers are publishing list after list of men going to Europe.

We must pray for peace. The bishop of Springfield (Episcopal) was here. He must be 80 years old and looks like Bishop Kemper. He spoke well on the spiritual significance of the occasion but spoiled it by roasting Emperor Wilhelm and discussing Panama, Colombia, etc. The prospect of our possible involvement in a war was depressing talk.

During an afternoon tea in the rectory after a series of conversations, the women were going on about President Wilson. They said if Chicago had the kind of support that T.

---

8. Refer to chapter notes, page 198: The Allen Survey

Woodrow Wilson gave to Princeton, they would be financially well off and more important, more democratic as to fairness in the academic hierarchy. The women were discussing Mrs. Wilson's death in August, saying they heard it was probably due to her despair over having tried to give him a son and the difficulty in bearing three girls. It was said as though common knowledge, which I had not heard previously and can't imagine. Ellen Wilson's girls are lovely young ladies.

H. B. M.

∞

Saturday, October 13

Dear Dr. Birge,

The books and net arrived on October 8. Zeleny is very impressed by the net and wants to know if they are available for sale anywhere. He asked me to review Woltereck's papers for a Journal Club on December 3.

I have Woltereck's first paper. His second paper can not be found at Illinois. The third paper was reviewed by Juday in *Science*, but Juday said nothing about "food," although it seemed the paper must have had something on this. Ask him.

Zeleny also asked me to review one of your papers and disagreed with the extent to which I praised your work.

I enclose a paper that has come out in the *Transactions* of the American Microscopical Society. I could not find it here at Illinois. I am interested in kinetic energy and want to know where and when it was activated by the "animals." Once again, shouldn't it be worthwhile to send me the thermocline? I have no books with me and it is a handicap not to have the things I am used to.

I will write again when I get unpacked. I am getting pretty good Cladocera, but they do not live. Lab is irregularly heated. I have never seen such a fall. I thought it was because we were further south. No frost and all flowers in blossom and trees hardly turned. I wish I had time to take the interurban to Decatur and Springfield.

Very truly yours,

H. B. M.

October 14, 1914

Dear Dr. Birge:

Rainy and cold today and no fire. I have taken a little cold and feel pretty mean. I felt this morning, when I crawled over for an eight o'clock quiz, that I could never get there. Feel a little better now.

Of the papers I am sending, the first is the one I am to report on, but I ought to have a general understanding of Woltereck's papers. I have not found the second paper here. They have some copies of *Hydrobiol*, but not all. I will look further. Librarian thought they had discontinued it (or at bindery). Will you see if there are later papers? I seem to remember some. They are not in library catalogue if here. I have not looked much further yet.

In the third paper Juday reviewed in *Science*, he says nothing about "food," and it seems as if that paper must have had something more. Ask him. I looked at the Duetsch paper a little but did not find anything.

I try not to go up and down stairs and lab is on second floor and the library on first. Dr. Ward gave me a key to Cladocera, so I will return yours in the near future. Smith has nets, but they are very old-fashioned. I told him you had a man who makes them but did not know price.

Yours truly,

H. B. M.

∞

October 15, 1914

Dear Roger,

This to let you know that I am moving to a boarding house, directly across from the Natural History Building, which is closer to the lab. The problem with my first house was that it was further away and 45 minutes at noon are not enough for lunch and the long walk for me, especially when the food is served at a slow pace. At Mrs. Baynes' house, the meat was always tasteless, floating in some indescribable gravy, beans and eggplant and seldom potatoes (are they expensive?). I hope Mrs. Baynes doesn't have a fit that I am moving out.

I am now situated at the new boarding house. It is handy for me. I sit at a table with Mr. and Mrs. Paul and am usually accompanied by an American Literature teacher as well as those teaching German and English. There is, at most dinners, a Miss Lupinski who is a librarian. She is thin and dresses in a sort of Aubrey Beardsley style with a narrow headband stretched tightly across her forehead. It seems to keep her thoughts enclosed for she scarcely expresses any. She goes by the name of "Sky" and when she does respond, her dark brooding eyes seem to precede a storm that breaks out of her cloudy persona, dampening the usually sunny atmosphere at the table. The only interesting item she wears are ropes of tropical seed and shell beads, which look almost like the long strands of beads that curtain the boarding house dining room from the entrance hall. I thought to compare her cataloguing system with the one I initiated at the Milwaukee Public Museum, the Dewey Decimal, but let it go.

I enjoy your notes. Don't know how to keep all of my mail. I throw out a lot every move I make. My love to you please keep.

Hattie

P.S. Did I tell you McGilvrey is here? I know him from Cornell when I was there with Comstock. He could hardly read or write—no degree—almost illiterate. I don't know if he has improved.

*Merrill often wrote more than one letter a day if news came up that she felt she should write or forgot in a letter.*

∞

October 17, Saturday

Dear Professor Birge:

I am leaving this house and moving to one directly across from the Natural History Building to try once more. The problem with the first house was that it was three blocks away, and the 45 minutes at noon was not enough for eating and the walk each way, especially when the food was usually served late. Furthermore, the quality of the food was poor—oatmeal and raisins, tender but tasteless meat of some kind floating in gravy, green string beans and eggplant, seldom any potatoes (are they expensive?), but rather good coffee and cornstarch pudding.

The table companions, except for Miss Simpson the librarian, are pleasant, mostly students at the new place. No. 610 South Mathews Street.

They called a meeting of teachers of freshmen last night which was rather funny. Babcock presided. He has only been here about a year. He has a weak chin, but strong delivery, although Meyer, his assistant, is much better. I suppose you know them both.

Babcock was very frank and said we wanted enthusiasm and scholarship from pupils, but the fact was that in the large and popular classes scholarship was poor, and that where scholarship was good, much work was required. Zoology evidently has a name for stiff work, for they joked a lot about it and even when the professors were giving their methods for having classes begin on time.

You will not hear from me again until I get settled.

H. B. M.

1010 1/2 W. California, Urbana

October 19, 1914

Dear Dr. Birge:

Professor Zeleny is both a pleasant and queer combination. He never looks properly shaved and his black eyebrows are as bushy as ordinary mustaches. With all that, he has the most girlish dimples and softest voice. His lectures are rather elementary. He has microscopical slides for me to identify and some ostracods sealed for five years but alive in specimen bottles.

As for Henry Ward, he still pleases my soul, he is such a disciplinarian. He gives splendid lectures to the elementary classes. When he was out of town, he had his parasitology class, of which I am a member, do library cards. I have done 40-50 up to now. Ward asked if I had ever done any library assistant work, because my cards were quite professional. I have worked on the Dewey Decimal System and had much cataloguing experience at the Milwaukee Public Museum. Meanwhile, Zeleny has me correcting freshman papers.

H. B. M.

∞

October 24, 1914

Dear Dr. Birge,

I am writing this soon because I have a little time, and because I have some things to talk over. The box you took the trouble to send was the wrong one. Do you not remember that there was no wooden box handy, and that I hardly knew how large a box I needed, so I put the things that were to stay into permanent wooden boxes, and left notebooks, etc. in something temporary, I can't remember just what. I think a pasteboard box. Anyway those things were on top. I have hated to bother you about them again. The funny thing is that the box you sent was the one that contained things I thought I could not possibly want. I put all that leech material there. I shall probably never use it again, but I hated to throw it away. Then I had some butterfly literature I used for South American stuff. I thought when I packed it that it was at the very bottom and was one of the things I might be glad to have. Well it is all on top now. I seem to see a way of attending to the stuff myself.

Illinois plays Wisconsin at Madison during the Thanksgiving recess. I understand rates will be $5.50 for return trip. The time will probably be short, Friday to Monday perhaps. Moths have gotten into some of my bedding, and bedding is so filthy here. I will need fresh supplies from there. Well this is something of my plan. I will write about it later. The Thanksgiving recess is from Wednesday noon until Monday noon, according to catalogue, so classes would not interfere. I hate the journey and change of heat but I feel pretty well today and ready to undertake it.

I understand now that the game is the week before Thanksgiving. In that case Ward may object. He would not let one of the boys off last Saturday although I was to take his work. I will see you as soon as I can.

Your notes came this morning, 10/16/14, and I will send this, posting it about three p.m.—as the next days are busy ones and I may not have much time.

Yours,

H. B. M.

∞

October 27, 1914

Dear Dr. Birge,

I saw Ethel Sabin when I was at dinner at the Thetas and was a little surprised to have her (Ethel) tell how Van Hise called the faculty together and stammered and stuttered. She mimics him well and told them they must learn how to speak in public. I added how he spoke on conversation and tried to quote "As panteth, etc." and could not remember what part of his anatomy panted nor what it panted after. We had an interested audience. I thought all the Sabins adored V.H. *[Van Hise]*

Mrs. James died Friday, and funeral services will be held Monday at three in the auditorium, no classes all day. I hope the faculty won't be ordered out in attendance. I think I won't go anyway as I know no one. The place is so big, I don't see how they will fill it otherwise. The students do not expect to go. Students behaved well after the football game. Word seemed to be passed round that it would be out of place to have demonstrations and the place was quiet. I do not know how much private celebration there was. The boys estimate there were 10,000 guests in town judging by the number of tickets sold. Students do not know how to mass. The cheering at the game was feeble. I was in the lab, and it was so perfunctory, I thought Illinois had lost. The boys themselves said there were not enough cheerleaders that people did not know when to cheer, etc.

I have not accomplished much this vacation. I spent all day Saturday sorting Cladocera and fixing Aquaria. Zeleny has an awful looking tank that has lots of *Moina rectirostis* some males, and I spent the day sorting them out. I shall be glad of your algae, but do not hurry about sending it. I will send some pickled Cladocera someday to see if I have them named right. Have you seen Sar's paper on variations in *Daphnia carina?*

Ward, head of the department of zoology, evidently expected me to "fish or cut bait" for he wants my preliminary exam in May, and French and German in April. I thought by that time I should be either quite dead or able to do as much as I ever shall.

Advanced students hate Illinois advertising as much as I do that of Wisconsin. University of Illinois had a "tag day" [*soliciting for donors*]. The Belgians expect to send their own car. Ward spent lecture hour boosting it and roasting someone who wrote to the *Illinois Journal* objecting to the "holdup." I did not run into the "taggers" and certainly did not run after them. Illinois is also trying to arrange a football game with Harvard for the Belgians and really advertising.

I am about sure I shall not come to Madison the 21st as rates are $5.00 each way.

H. B. M.


610 Mathews Street,

Sunday, November 1, 1914

Dear Dr. Birge:

I did not go to James' convocation, and am half sorry as part of it was funny. It seems that some society asked President James to give the talk, which, by the way, the eastern graduate students criticized as "a freshman stunt." Evidently James wanted to give it. The funny things came from the president of the society, who began by saying something to the affect "that it was well to develop our own resources for home talent in the shape of James," and went on to say that "even if James was not much of a great educator he was a good president." Trust the undergraduate to sense the limitations of the faculty.

This town went crazy over its Minnesota football victory. The student body met the team at 12 today. I do not know what the pious ones did. Wisconsin did rather well to tie Chicago to no score. I told you I am with the undergraduates now. From faculty down, this institution hates C.U. No one tries to conceal it.

I may send you James' speech. I have it and do not consider it very good. It may amuse you, although probably you know the substance by heart.

H. B. M.

November 6, 1914

Dear Dr. Birge:

Yours of Nov. 4 received. I got a little panicky about the mail. The girls said that we would be obliged to go to the post office for it. There was something about renumbering the streets in Champaign, but so far my mail has been prompt and well delivered. The street is South Mathews 610. I sent you a letter Saturday, the 31st, in the afternoon. I was too late for the 12:30 carrier, but you said you got it Saturday which was good time. My mail is left in the box at the house until I take it out. Mail sent to room 310 Natural History Hall will always be delivered as there is a box there.

Do not bother about your N.Y. paper. I have Zeleny's and that is enough. I will write later as to points to be made. I want to give you a good review. I shall try and see Ward before I post this as to whether I can come to Madison. I fear rates will not be as good as I expected. A week from Saturday is the great day when Illinois plays Chicago here. I will not say anything about your Lakes paper until I know whether I am to talk with you. I fear not, for I know the journey will be a hard one.

I am sorry you are so driven with administrative work, for one can't do literary work without leisure. I wish success to you on the Stevenson paper, however.

You were considerate to tell me your political guess was 50 percent right, and not tell which 50. You can't buy a paper here, and these boys are from everywhere and don't know or care. I saw this morning a list of governors elected naming Phillips, and therefore suppose Francis McGovern was elected. What a blow for LaFollette, who is to, or has, lectured here. Do tell me the news.

Ritter, California was here Wednesday and we talked. I was quite interested. He has a man now who is working on a large scale on a certain species of wild mice that extend from Alaska to San Diego. Crossing them, studying environment, adaptation, etc. Thoroughly fascinating. I see that he got very definite results from his ocean work and he has quantities of data. Smith asks where he can get your net. I said from you.

Very truly yours,

H. B. M.

610 South Mathews St.
Urbana

November 15, 1914

Dear Dr. Birge:

Not much to write, but I will send a line in reply to yours of Nov. 12. I should be glad to read your Stevenson paper. You will have some time at Thanksgiving will you not, or are you so forehanded you expect to have it done by that time? I am counting on that time to review your paper as you asked. I think I have all the literature. The plankton Cladocera I have not read. The book was too large to lug home. I have it here and will get at it tonight. I understand that comes first, suggesting the other problems. I thought to take up the thermocline and conclude with it (as the review of the N.Y. lakes) the real paper you asked me to review. By the way, my copy came Friday. Of course I do not review the work of your assistants by name and the sight of Wisconsin Academy and all the rest makes me so mad, I shall be as uncircumspect as Smith. I certainly miss no chance to roast V.H. [*Van Hise*] with the undergraduate, and Wisconsin methods. Kahlenberg is held in scorn, and it is quite fun to give examples of freshman chemistry as he teaches it.

Trottman won't have a chance to be cantankerous much longer will he or has V. H. got the regents fixed so Phillips can't appoint anyone? I hope if he has, the law will be changed and P. will get as much of a chance as LaFollette did. Was the convention of the governors, McGovern's last blow out? I expect each letter to hear he is senator by a vote or two.

Why is Wisconsin sending philosophy students away? Ethel Sabin is still here, said McGilvrey advised her to come to Bode, I suppose for her Ph.D. as she took her master's degree at Wisconsin last year. Carl Haessler is here, assistant in philosophy with an A.B. degree. Don't the Oxford people get any degree?

I am glad to know about Cobb. I have sort of a feeling that I ousted Miss Cobb. She was assistant in zoology last year, and is assistant in education this year, but hangs around. She asked me if she could assist in my lab Saturday mornings (one day a week). My co-lab worker told me to ask Van Cleave, and V. C. and Ward seemed to turn her down. She wanted to take one hour with Zeleny and he refused. She did a paper with him last year on inheritance of mental traits using Courtis tests on faculty children. No one seems to want to publish it. I mean to look it up. Things seem pleasant enough. She comes to Journal Club, etc. I thought you might know her personally. She seemed to know you, at least said you were writing for Ward as her father was also.

Can you without trouble tell me if Trinity College, Connecticut, is denominational and what denomination? It used to have another name—can't think what it was.

Very truly yours,

H. B. M.

*Merrill's letter, typewritten on departmental letterhead, shows that the zoology faculty consisted of H. B. Ward, in charge, J. S. Kingsley, F. Smith, C. Zeleny, V. E. Shelford, H. J. Van Cleave and Harriet B. Merrill, assistant.*

*During this period Merrill wrote regularly on Fridays and Sundays, the latter mostly from the Natural History Building.*

```
        FACULTY                                                    ASSISTANTS
HENRY BALDWIN WARD, IN CHARGE    THE UNIVERSITY OF ILLINOIS    BESSIE R. GREEN      H. V. HEIMBURGER
J. STERLING KINGSLEY               DEPARTMENT OF ZOOLOGY       H. G. MAY            HARRIET B. MERRILL
FRANK SMITH                              1914--15              R. H. LINKINS        T. B. MAGATH
CHARLES ZELENY                                                 J. L. CONEL          GEO. M. HIGGINS
VICTOR ERNEST SHELFORD                                         H. E. CHENOWETH      H. F. METCALF
HARLEY JONES VAN CLEAVE                                              RACHEL BAUMGARTNER
                                                          C. W. REDWOOD, SCIENTIFIC ARTIST
                                                             W. MATTOON, STOREKEEPER
```

November 19, 1914

Dear Dr. Birge,

Freshmen had a grand quiz, mid-semester, this morning. Papers numbered, not named, so as to get fair marks. I was told we were all to mark, but have no papers as yet. My classes ought to do well, but I have half a dozen colored people, and one Chinese, etc.

There is also a conference of high school teachers. I went in for a minute to hear a youth of 20, who had not had any science, tell how he loved the general science courses in the grades. I was too wrathy to listen any further. The rot grows worse every year.

I hate to have Shelford come to direct the lab course after Thanksgiving. His fool work draws a number of students. They go out to this ditch and Shelford and all the rest forget the nets. One of the girls has mine and so they strain a liter of water, come back to the lab and count the beasts—get average of 25 to liter and call that "quantitative biology!"

If you can, after the Stevenson paper, look at a paper by Shelford and Allee in *Science* Vol. XXXVI, July 19, 1912, on $CO_2$ effect on fishes.

I feel better. I do not know whether my heart has given a notch or not. I can breathe fairly well and eat better than in a long time and sleep well. But the living conditions are worse than South America. I never knew anything worse. I know the food is dirty, and spoiled, but I eat it with relish and probably shall enjoy never taking a bath. The clean sheets once a month are dumped in the room, never put on the bed. We have little heat in the house either.

We had considerable excitement yesterday. I looked out at the neighboring house, about 20 feet away, and saw smoke at 6 a.m. and wished I lived there. Yesterday morning I gazed out and found the roof blazing also. Fire was controlled and when I reached the lab someone asked me how the Thetas were getting on. It seems they took fire a little later and the third story went.

They have rented a house in Urbana. What troubles me worse is that we will never have any furnace fire now. It actually has not been very cold as yet, fortunately.

Yours,

H. B. Merrill

610 S. Mathews St.

November 20, 1914

Dear Dr. Birge:

I am taking good 11 a.m. time to get a note to you acknowledging the book on lakes. Illinois has the survey publications, and I had a copy, which seems a straight quiz crib from Fenneman. A little about the tables seems worthwhile and that is all. I was glad to get the outline and shall have a bigger job than I thought in reviewing your last paper. I may not give it orally after all, although I think there is little doubt that I shall. It seems the Journal Club is larger than I thought but all do not come. We have had two holidays, etc., and some attending were guests of members. I am to hand in a written report the third of December and report orally if requested. The J. Club can be an awful bore for there is no discussion, and anyone with eyes could read the original paper in a third the time some graduate student is blundering over it. The physiology Journal Club is good. It is very small and the profs. discuss and do it well. Took up the paper of Charles Bardeen and Dr. W. S. Middleton (clinical instructor of medicine and student health services at the University of Wisconsin in 1914) on the effect in afterlife of the over-strain of athletic champions, etc. You know what I mean. I am able to go to that session regularly now.

What shall I call your work? A catchy title will help to get it a public review. Zeleny suggested "Ecology" and I snorted, and he laughed. The meetings of the Journal Club are public and are printed in the weekly bulletin so it cannot be too long.

*Her tentative title proposed near the end of the semester was* Studies of the Physical and Chemical Characteristics of Inland Lakes (freshwater) and Their Effect on the Life of the Lake. *Although, she disagreed with most of Shelford's ecology, her report's emphasis was ecological.*

This institution is queer. The storekeeper came and gave us a dozen or so sheets of this paper but no envelopes. As nearly as I can guess, the woman who does not clean my room, stole my supply of stamped envelopes! I do not know what this will go in.

Yours,

H. B. M.

*Harriet Merrill's Journal Club topic was changed again from a review of the thermocline paper, published ten years earlier (Birge 1904), to the paper on the Finger Lakes of New York, which Merrill had completed and had just been published with credits to Birge and Juday (1914). She had Zeleny's copy of the New York paper and did not need a copy from Birge. She realized that reviewing the paper would be a bigger job than she had anticipated. A written report would be due on December 3, and then she would know in a week whether or not she had to present it orally.*

610 South Mathews Street

November 22, 1914

Dear Dr. Birge:

I came up to the lab for books, and partly I suppose because I have a key, and it is sort of reflex action to spend an hour or two at the lab on Sunday. It is the one place on any campus where you can have "a quiet hour." Even Wisconsin is calm but this place is peaceful. Someone told me there was an historical play at the movies and I asked about it on Saturday. Saying, "What play is being given today or tomorrow?" not remembering that it was Saturday, and one of the boys said very promptly, "The only place you can go to tomorrow in this town is church."

The boys were rather quiet yesterday for the successful end of the football season. The Wisconsin game must have been quite spectacular. Perhaps they were quiet because of the newspaper yarn that a city merchant had absconded with the stakes he held on football bets, then told reporters the boys had looted his store—blaming them. One of the boys said, "I am a senior and have never seen the place, it is so insignificant." I think I shall ask some of the boys who told me they had bet on the game in order to determine who was telling the truth.

I went to the first and only function I have attended here so far to hear Kingsley talk to the graduate club on Saturday night. Dean Kingsley, University of Illinois science faculty, spotted me. He really does know me so I concluded it was best to be seen. I like him very much.

I had a very good time for I had a long talk with the Burges, who are the most interesting people I have met. Mrs. Burge taught for sometime with Miss Clapp at Mt. Holyoke. She made a very good story of how when Miss Clapp was at Chicago, and Donaldson was ill and had a German substitute, the substitute came to class one day and began, "You do not know anything. I will tell you. You will write it down," and he began: "Such a nerve has two branches, etc. You will write it down!" Miss Clapp felt they certainly were getting worse than nothing from Donaldson in comparison with the German prof.

Kingsley's talk was pretty good. He compared the ideals of English and German universities, and said that in a way America was not getting the best of either, and went on to develop the idea that a Ph.D. should mean training, scholarship, creative research, and the last most intangible thing of all, which was of the spirit only—culture. The president of the club introduced the Dean by saying that a Ph.D. from Illinois would be worth more 20 years to come than it is now. These students are nothing if not frank. It pleased my soul to have Kingsley explain that library students at Illinois were not and could not be in the Ph.D. class—considerable frankness on all sides! Miss Simpson (library faculty) was there and so was the cooking school. Miss Simpson was slightly subdued, however, but no more of this for me. Miss Simpson gave me some 15 minutes of her valuable time but I do not go to the Graduate Club to meet the trade school.

I am writing what I think of first, and it may amuse you to know that Mrs. Burge inquired about Ted, and as nearly as I got it, she started to tell the story, and then thought it did not sound well, I imagine. But the Johns Hopkins people got her engaged to Ted instead of Dr. Burge! I think they all worked there.

I think I am a little too much mixed up with fires. I think I wrote the house next door caught fire, later in the day the Theta house and still later the woman who took my room on Cal. Street, and was dissatisfied and moved, was burned out the day after she moved in.

Perhaps this is gossip enough, I do not seem to have anything very important to say.

I do not doubt that your Stevenson paper will be good. Who is to discuss it? You ought to be as clever as Slaughter in getting it well placed.

Very truly yours,

H. B. Merrill

∞

November 23, 1914

610 South Mathews Street

Dear Dr. Birge:

J. Kingsley is very pleasant—asked if I was at Wisconsin when he was. He said that Germany was in danger of Civil War and they had entirely misjudged the situation in taking big chances—the emperor was forced into the position. I told him about Vilas's troubles and he said, when he got his "ticket" there were 12 chair positions to choose from and Vilas did not know how to proceed. A young fellow from Granell says G.'s president, Mead, was offered the presidency of both U. of N. Y. and Johns Hopkins. A woman in chemistry here says Ruth Marshall failed to get the Progressive Party to nominate her for county superintendent. She took exams in Chicago and has been assigned a substitute place in English to teach the Jews and Negroes.

The woman in chemistry took her degree at Columbia and says that Frank Smith has been able to get everything he wanted and has fitted up an elaborate laboratory. This does not set well with the other departments. President James, University of Illinois, recognized several departments who were loudly applauded. He mentioned Ward's parasites to no applause. A student in physiology claims he has fights within his department. Has Vassar a president yet? Where is Y? Professor Paul says Raymond Rohlins will not be elected senator in Illinois.

Yours truly,

H. B. M.

∞

November 25, 1914

Dear Professor Birge:

I have written a review of your papers, and have sent you a copy last week for your comments. I received the proof and am concerned that the paper on Stevenson and mine on the New York lakes might have gone astray because it was not well packed.
Please let me know.

H. B. M.

*Two letters were mailed during this period, and in addition the one of December 8 written on her return to the boarding house is concerned entirely with the hospital. "Cold and starvation and dirt nearly put an end to me. The doctor thought I had symptoms of pneumonia. My fever and pulse were both about 100 when admitted, but they came down rapidly. Dr. Newcombe said I had a very narrow escape from pneumonia. He did not hesitate to use medicine and hot water bags and hot oil and camphor and about all the dopes I ever heard of—throat spray, —flannel on the chest, etc.," She could not talk at all during this period and hence could not think of classes.*

*This letter was originally three pages in length, it has been cut to the essentials. To include all letters in their entirety would have been a book in itself. Letters are deciphered from the originals which are filed at the University of Wisconsin. Family letters are in the M. L. Hartridge collection.*


December 5, 1914

Dear Dr. Birge:

Your Finger Lakes papers came safely. I will take care of them and am reading them here. I think I am going to be much better. It has rained continuously all week and does not stop day or night. I want to post this note right away. Hope you are getting out of your troubles too.

The Wards have been very nice. They sent me three dozen American Beauty roses—makes me uncomfortable, though much appreciated.

    H. B. M.

∞

December 8, 1914

Dear Dr. Birge:

Received your note Saturday. It was cold in the lab or I would have written there.

I am much better today and wonder how long it will last! My present doctor treats me differently from any I have had. He has given me pine tar for my cough and it seems to get rid of the mucus. He says I have considerable bronchial and asthmatic trouble. I have no fever but feel as though I have when under pressure.

I have enjoyed reading your paper very much. I can see unfinished parts, but I think you could shorten it and it might make an interesting magazine article. I will return it in two days—hesitate putting it in the Christmas mail.

This is just a note to let you know I am on deck.

    H. B. M.

610 South Mathews Street

December 12, 1914

Dear Dr. Birge:

I seem to keep up a sort of "off again, on again, gone again Finnigan" performance. I am feeling pretty well. Do not have to get Journal Club paper in until after Christmas. I feel like going to some hospital or sanitarium for the vacation. I do not think I shall try to go to Madison; i.e., shall not go to the University of Wisconsin at all.

My classes did pretty well. Van Cleave congratulated me on my student's high marks, but perhaps it was to console me for low ones. I had one 100, twelve 90's, and about a dozen who failed. About six I had reported as due to fail.

I wish I liked and trusted Shelford more. He has his "set" that he is pushing. The trouble for him is they are youthful and indiscreet and say he is thoroughly disgusted with the general slackness of things. I never found Chicago very severe. I have one of their junior college men who takes lectures and quizzes only and who is one of my people who was 38th in exam. I am not bothering over any of this and think I'll keep my mouth shut.

Some of my Aquaria are doing very well.

It is 5:30 and I must go but will add a few lines. It was four below zero here last night. Flag up for fair and warmer. I did not go out at all on Sunday and stayed in bed a great deal of the time. I am busy catching up and shall not expect to write nor to hear from you until things ease up. Shelford improves. He has his own way to make.

I had not realized the Allen matter was so serious. McGovern was very cheerful in getting the University into the mess. And I suppose V.H. is equally cheerful in letting you pull him out. His geological mind works so slowly he would never know what Allen was driving at. I am sorry it has been so hard on you, and so worthless.

Miss Sanford finds some difference between going six or seven blocks rather than across the street, I thought she would.

I am glad Eugene is doing so well. Mrs. Trelease called on me at the hospital and said William had to have a drainage tube and that every day he had his bandage removed (peritonitis). *[This was a serious condition according to medical statistics, a ruptured appendix was considered fatal up to 1950. Merrill's older brother died of the same occurrence just before her own death.]*

Very truly yours,

H. Merrill

*Brilliant birds and giant butterflies flew unwarily about in the symbiotic sanctuary of Sao Paulo Botanical Gardens. I expected to see such magnificent specimens confined. I was enchanted — and the only species captured while in the "conservatory with no bounds," was me!*

*Harriet Bell Merrill*

December 14   5 p.m.

Dear Dr. Birge:

A good cool, clear evening for your paper. I know it will go off well—and shall hope to see it sometime.

∞

610 S. Mathews St.

December 28, 1914

Dear Dr. Birge:

Your note came as I was on my way to the lab, so I will write just a line now and more later. This is the first day I have been over here. It has been, I don't know what, -12 degrees?, and I simply went to bed. My landlady has been good to me, brought all my meals, and tended me. My cough has about gone but I seem a little asthmatic and weak in the knees, but if the cough goes, I ought to be about normal.

I have no end of work. I am now trying to get an outline of Inland Lakes. I can't even get the beginning I sent you. I think there is just a word wrong in one paper. The T. and its B., Sig. page 10, middle paragraph, tenth line from bottom, should read windward not leeward, "nicht wahr?"

The Slaughter case seems strange does it not? Too bad. I suppose Gene is all right, as you have not mentioned him lately. Miss S. was now gone nearly a month. I should think it would show you that you ought to have an "understudy."

Mrs. Carson is wonderful. I know her old house must have frozen solid, however. I will send this and write more when I send outline. I feel better and have done considerable work today.

Very truly,

H. B. M.

P.S. The cold weather has started with its attendant problems. I went home this noon, not intending to go to my room, but the girls told me the plumbing pipes had burst for the third time since I have been living in the house and my room and clothes closet flood every time. At times like this I think of "Rolling to Rio," for the third and last time.

610 S. Mathews St.,

January 2, 1915

Dear Dr. Birge:

I haven't much to say, but will write just a line. What do the Chicago papers mean by saying Taft has been asked to take the presidency of U.W.? The regents do not change do they? Is V. H. *[Van Hise]* getting ready for a political job, too?

I suppose you know what professors get here, but did you know Ward gets $5,000 and Zeleny $3,700. President only gets $6,000 I think. W. *[Ward]* will not be leaving here very soon. Students say Z. *[Zeleny]* is a very poor lecturer to elementary classes. He has a course in heredity. His upper class students are not very enthusiastic. Shelford seems rather popular, but he is so superficial.

Don't worry over my review of the Survey papers. I can get them out if I ever get time. You see I have been laid up since Thanksgiving and with three undergraduate classes I have barely kept going.

I have such quantities of notes of my own to get to. I hate lecture courses. I could read their stuff in half the time it takes to copy, but Ward is daffy on notes.

Another bit of gossip, James announces the engagement of his sick, lame, hip-diseased daughter to Fraser who was an accountant at U.W.

I have been feeling quite sorry for her, I thought she was so helpless.

H. B. M.

∞

January 4, 1915

Dear Dr. Birge:

Your note and outline came today. I am sorry you worked at it when you felt ill. Does the political change make you nervous, or are there other worries? Is it the worry that seems to set your insides off? My cough is gone but I'm very short of breath. I did all the work I could, but the furnace didn't work until noon. I have my class notebooks well marked, my own notes typewritten and a dozen pages for you on the N.Y. Lakes. Am letting you know now as next three weeks will be hard.

I had a letter from Katherine Paine telling me her mother died at Christmas. That breaks up the home.

Very truly yours,

H. B. M.

P.S. If a Miss King there at UW has inbred white rats for generations and the more inbred the bigger they grow (contrary to all previous statements), I think she ought to harness them up and determine the power of work they could do!

January 5, 1915

Dear Dr. Birge:

 I am writing in haste because I am taking an hour off to hear Shanan (Eng.) read poetry at 7:00. Received the proof this noon and and am much worried lest my review of N.Y. Finger Lakes went astray as it did not come back. I think there was nothing but the N.Y. Lakes of which I have a copy, but I should be sorry if you lose Stevenson. Your papers are not so bad. It's just that the cold of vacation left me blue as well as lazy. I do expect to give you a good review.

 H. B. M.

∞

January 8, 1915

Dear Professor Birge:

 I am sorry you felt obliged to write so much for me. Some of it was as you say repeated and "thermal resistance" was easy to state. I thought I must explain your table and the formula the math prof. gave. The trouble is, I have had to read your whole dozen books before I could begin the last two.
 Won't you state your law of thermal resistance in English? Write out the words. I have neither physics nor mathematics books to consult. Give example in figures. What is heat budget for—just a law? What becomes of heat in summer that isn't absorbed, etc.? I presented my paper at the Journal Club on January 22, and thought it went fairly well, although I was allowed less than 20 minutes. Shelford did not come. He might have with profit, according to his quotations of your work, but let him waste his time if he wishes.
 I am very busy, for of course, I get little consideration for feeling ill. I look and seem better but am weak and out of breath.
 Ward spoke of the Wisconsin investigation and I could imagine Zeleny said he saw Pearse and I could image him smelling around and saying "we" could get back at Allen now since he had published a preliminary report. I said Allen had been to classes and sent Normal people there, etc. and reported on number of hours laboratory was not in use, etc., etc. Is it in the papers? What ones? *[Refer to chapter notes 8, page 198: The Allen Survey.]*
 I know I have not written a reply to your letter but I have not had time to think. I am tired but must go to class at 3:00 and it is 2:30.
 I wrote of English reading. Did not compare with Pryer. Sherman read from *Lady Gregory* and some of his own fool stuff. He impersonates fairly well. Everybody hopes Dean Kinney will read *Barnie*.

 Very truly yours,

 H. Merrill

610 S. Mathews

January 12, 1915

Dear Professor Birge:

I never was quite so busy. I shall have to take two exams myself and mark a bunch of freshman papers. In addition, I am to give the review of your papers a week from Friday. Zeleny asked if I wished to wait for next semester but I said no. For one thing I fear they may cut it out and it is as fresh in my head as it will ever be. I did not write a good review— too conversational and it needed diagrams, but I can talk it. Zeleny asked for about a half hour of discussion. There has been very little discussion of papers thus far. Zeleny did not like the praise I gave to your theories. I felt the praise came out of our work plus his class work. He has been giving a lot about cell lineage, etc. and every experiment has been different. The boys criticized the work and said it was merely remembering a lot of disconnected things. I will go slowly with praise of your work.

We finally have enough water for baths. Feel like I have been hibernating. Glad you had some recreation in Milwaukee.

Ethel Sabin told me about Allen. She says Monroe goes to Princeton as a result. Probably would have anyway. Ward and Zeleny said Allen was dismissed from the Municipal Bureau at N.Y.U. You know of course?

Pres. James' daughter who was born with a congenital hip, went through University of Illinois, studied music abroad, sings a little, she fell while entering a carriage and was in the hospital. She is supposedly quite helpless. James has had many German professors here—apparently did not want to have more—and got into trouble over it.

In regard to your paper, I have one thing about it and will send you the review. They have taken away the typewriter so writing is not easy. I am studying the factors on pg. 997, line three (diagrams). Is this the explanation?

News in the physiology department—they told of an apparently normal female child of 12, growing unwell for several months and there appeared an abnormal growth in the abdomen. An operation showed it to be a developing embryo. The child thrived after the surgery. Burge is working on cataracts and analyzing lenses, finding abnormal amounts of calcium, put fragments in chemical solutions, subjected them to ultraviolet light rays and got cataract effects. *[results illegible]*

Yours truly,

H. B. M.

January 13, 1915

Dear Dr. Birge:

Shelford is bound to disagree with your theory over $Co_2$ and I must be ready for him. He has fixed troughs and larger fishes and claims $Co_2$ is toxic to the larger—that they will move away from 5 to 30 cc in excess of what they are accustomed to. I should think it would give them "hiccups!" Will open vessels hold 30 cc in solution? You say that $Co_2$ will be lost to air if much in the height is dissolved. Shelford says this is the case in measuring seawater when the tide is coming in and out. That was questioned? If you have time to read the Wilson paper and earlier Batemar he refers to, I would like your opinion on his hydra story. I read this in library while waiting for Trelease. My paper on lakes *[New York Finger Lakes—Merrill's own research in conjunction with Birge's views.]* has been published. One of the pupils expecting a Ph.D. referred to it.

Yours truly,

H. B. M.

∞

January 15, 1915

Dear Dr. Birge:

I am sending on my work on the N.Y. Finger Lakes of which I have a copy—do not return—but I should be sorry to be the cause of your losing Stevenson.

Maybe it is the impatience that comes with age but nowadays I hate lecture courses. I could read their stuff in half the time it takes to copy but Ward is daffy on notes. When he has to pad a lecture with kinds of muscles of the heart, and no lab work—all lecture—I am bored.

I shall certainly die of parasitology. Ward told all about shell glands and vitelline glands and oviducts and then said it was all wrong. I never saw a parasite other than those in Madison boarding houses and why should I care if certain scales are useful in determining species? They teach crayfish according to Husley and fix up gills to match. I have 17 gills—you do not remember I suppose.

Yours,

H. B. M.

*In spite of her physical disability during this period Merrill was still anxious about the quiz sections she had missed. She had been told how her students had made out in the examinations. Van Cleave congratulated her, although she did not think she had done any better than the other instructors. Although she had been in bed on January 17 and the two previous days, she wrote, "I hope to get to the lab tomorrow, but not to stay. I have four recitation quiz classes and shall try to get through with them. This is the third time I have had to give up classes."*

*In this letter of January 17 she suspected that Ward might ask her to step out, as one of his girls, who was having trouble with her eyes, remarked that graduate school was no place for invalids.*


January 17, 1915

Dear Dr. Birge:

I wonder if I have come to the end of my rope. I had to give up again Friday. The fluid accumulated so I could not drag myself upstairs. The Dr. says I must rest for a time. He says my heart is not chronically bad, but there is some regurgitation and that it will continue, if I keep on.

I have been in bed for three days now and feel better…I don't know what to do. I could stay here and work with Zeleny as I am able, I suppose. Zeleny is anxious to have me work on pure lines. He brought interesting reports from Jenning's lab in Philadelphia.

Yours,

H. B. M.


*On January 23, the day after her Journal Club report, she felt "some better and shall get through term," but on January 26 her entire letter to Birge was the following:*


January 26, 1915

Dear Dr. Birge:

I have gone pretty much to pieces again. Sunday it was 18 degrees below. Milk froze in my room and did not thaw all day and I did not get a mouthful to eat until two p.m. Today I am staying in bed and shall not do much of anything more. Dr. Newcombe wants me to go south, but I think the journey too expensive. He says I ought to be able to work by fall. I don't know. I think I wrote you that the house next door caught fire? Then, later in the day, the Theta house caught fire and still later the woman who took my room on Cal. Street, and was dissatisfied and moved, was burned out the day after moving!

Perhaps this is gossip enough, I do not seem to have anything very important to say.

I do not doubt that your Stevenson paper will be good. Who is to discuss it? You ought to be as clever as Slaughter in getting it well placed.

Very truly yours,

H. B. Merrill

∞

# Using Some of the Last Good Light
*[A term Merrill often used in her photography work.]*

Jan. 27, 1915

My Dear Marie,

I was told that you were here to see me and I was too tired to respond. The medication they give me to sleep causes me to be occupied all night in dreams. I am either working in a lab or having endless conversations with an array of persons—some very real and known—others odd and strange to me. I often awaken simply exhausted.

There is nothing more stirring than a concert playing John Phillip Sousa marches. I don't know where we were but you and father and I were in a park when suddenly, everyone was scurrying about shouting, "The president has been shot!" When I woke up, father was holding on to one of my hands and you the other and I would not let go. Was that when you were here?

The other night I was in a consulate dining room. A housemaid was wrapping the best silver in cloths with camphor to prevent the extreme tropical humidity from tarnishing the pieces between their usage. I happened to see my reflection in an ornate mirror above a console. It was eerie. The silver backed mirror, cloudy with tarnish showed a tattered image, while elegant young women in their high heeled slippers clicked past me on the marble floor in the foyer. I was awakened by a nurse who had placed a poultice, heavy with the scent of eucalyptus, on my chest.

I am certain that the clarity of my dreams of the beautiful orderly gardens planted by the English, the botanical gardens of South America, the tropical rain forests and our own deep, pine fragrant forests would not be imprinted on my mind so vividly if I had not focused on them through my camera. I recall more than one occasion back in "Eden," trying to get my photographic gear dried out from the humidity and I still hope to catch some of the last good light on certain subjects of importance. My dreams sometime soften the often negative images in real life and I am embarrassed to admit that there are times when I would rather not be disturbed. I find that is a condition that comes with confinement.

I am hoping that one evening I might be back in the ballroom of a governor's mansion, attired in such a way that I may be asked to dance. I never had the desire to glide across a floor. It can't compare with ice skating but should Dr. Cruz ask to give me a turn about the floor and Señor López, handsome in his uniform, ask to cut in, that would be a dream! I would hope not to be awakened until the musicians stop playing or Gabriel plays his solo!

Forgive my prattling on but memories are a comfort now. Seeing my family is of course the ultimate of pleasant diversions for which I demand to be told to respond if they are present. Above all, I have a sincere urge of late to make it clear to the family…although there was a period when we were surfeited with Friedrich Nietzsche's credo, on the UW campus, I never succumbed. Roger seemed worried about my soul.

Love to you all,

Your old Hattie Bell

∞

*On February 7, 1915, Roger suggested when he visited his sister Hattie in Champaign that she could live with Marie, Nate's wife, in Chicago. Even though Marie requested that she do so, Hattie wrote that she preferred to stay in Urbana.*

February 8, 1915

Dear Dr. Birge:

If I could get a room on the first floor and near, I would try it a little longer here doing part-time work. The girls tell me Ward said he never had anyone so highly recommended to him as I was. I do not know whether you were one who recommended me or not. Zeleny wants me to continue, even says I may work as I please, etc.

Dr. Newcombe says I ought to be able to work by fall. I don't know. I am thinking of several places—Oconomowoc, Stevens Point, etc. Dr. N. wants me to go south, and through the Paines I could get a place there at a reasonable price, but the journey might be too expensive. I shall post this now and will write when my plans mature.

Yours,

H. B. M.

*After Merrill was released from the hospital and spent only a few days in her very pleasant quarters in Champaign, Roger then ordered her to Milwaukee. He had arranged for a first floor room for her, which ended up being the third floor, and her condition deteriorated rapidly. Her time in these accommodations was very short. She left Champaign for Milwaukee on February 22nd and wrote on February 28th to Birge about the unsatisfactory accommodations and was admitted to Columbia Hospital the next day.*

*By February 10, she had greatly improved. "I am feeling so much better today—drugs (for the doctor does give me plenty of drugs) and rest, simple food and no stairs and I am quite myself. He says I ought to have some months of rest. I can't breathe at all when I climb stairs. It is the old 'loss compensation' and the doctor tells me an attack of indigestion would take me off."*

*She had a long talk with the doctor on that day. He stated that any resumption of continued work would do her no good. He did not want her to travel far, certainly not as far as Stevens Point or Ashland, but yet she had been thinking of going there for the remainder of the winter and then back to Mrs. Crummy's place in Oconomowoc for the summer.*


February 15, 1915

Dear Dr. Birge:

I was admitted to Burnham Hospital on January 29 and was released from there in Champaign after February 10 and have spent many satisfying days at 201 Park Avenue. I have city heat day and night, and the meals and both facilities are very acceptable. If I had living circumstances such as these all along, I wouldn't have had so many health complications.

I was happy to hear that Ward said he never had anyone as highly recommended to him as I am and Zeleny wants me to stay on. The Kingsleys and others have sent me flowers, unnecessary but nice of them to keep me in mind. It must have been difficult to have Garfield come when you were so busy. I hear news from Wisconsin people here that money for the University mens' dorms is to be withdrawn. I have always thought it wasteful. How are things coming on that anyway?

Dr. Zeleny tells me Frazer is engaged to the James girl. She did not want to stay in the University anyway and he preferred public work. This is all for now.

Yours truly,

H. B. M.

Milwaukee

February 22, 1915

Dear Marie,

    I have delayed sending this note as Roger has insisted that I come to Milwaukee. He had arranged for me to have a room on the ground floor but I have ended up on the third floor instead. The stress of moving has made my condition worse than ever. It looks now that Dr. Ogden wants me back at Columbia Hospital again.
    I can't wear a corset and have to have my breakfast brought up three flights. I just can't do anything. It is dreadful to be here—such an invalid.

    Yours

    Hattie Bell

∞

230 Farwell Avenue
Milwaukee

February 28, 1915

Dear Dr. Birge:

    I don't know why I did not get a letter to you Friday and I don't know what to say. I'm not quite myself in the hospital, but had to do a final moving of things in a hurry in order to get away and now I'm worse than ever.
    The swelling left my feet entirely.
    I have such a nice room. I will try to write occasionally.
    You do not regret the Trottman resignation—"nicht wahr?"
    Do you want the *Daphnia*? I can send on Ward's. I have it here or I can send the wooden box to you again from Milwaukee. Will you send me Cheney's address, a faculty man wants it.

    H. Merrill

Good Friday, April 1, 1915

Dear Dr. Birge:

    I do feel better but it grieves me to realize that I am out of my mind. You were here and I didn't dare to see you. What does "Dean Birge ends 40 years with the University of Wisconsin" mean? It was in the March 28th *Milwaukee Sentinel*. I thought I had written you since then, probably not.
    I am very tired and for that reason, sleeping well at night. I hope things are catching up for you.

    Sincerely yours,

    Harriet B. Merrill

∞

Columbia Hospital

April 3, 1915

Dear Cousin Nora:

    I don't really know how I am. I just got a note off to Dean Birge. I heard that he tried to see me. Few people are permitted to see me and they are giving me drugs to make me sleep. I feel slightly out of my head and think I am on a boat. Nurses are very nice and understand. I told Birge that he should probably enjoy the dinner party for him after all.
    Thank you for the lovely flowers and most of all for reading to me from Emily Dickenson. You know my lifelong love of books. You should have them. But, my eyes "focus" better now when they are closed. I look at the flowers and remember the gardens in Rio. The Avenue of Royal Palms, the giant ferns, the tropical butterflies, like masses of fluttering blossoms. All of the species seemed so very free in that symbiotic paradise. There were no wire boundaries in the garden conservatory. My hope is that the magnificent specimens can continue their escape—from being captured and put in cases in the museums. I keep trying to escape. I will fly with my boots on!

    Your loving cousin,

    Hattie

*As ill as she was, Merrill was still thinking about the Cladocera specimens and wanting to get Birge's back to him. Birge wrote her a letter telling her that he was to be given a dinner and hoped she could be there on March 17. She responded somewhat sardonically with "You may enjoy your dinner party after all" as though Birge had expressed reservations about the event in a previous letter to her.*

*Her last letter, dated April 8, weakly scrawled in several places, and difficult to decipher was the first letter in which she addresses Birge as "Dean," possibly because this was the office Birge had held since 1891.*

April 8, 1915

Dear Dean Birge:

I received the invitation to the dinner in your honor but was too ill to attend. I wrote you and sent word. I think I did both. I am interested in the work on the sun machine and I should have liked nothing better staying on with it at Illinois. I don't know how I am. Roger writes about Christian Science—supreme being, etc. which shows what he is thinking. I can't say anything, it is so queer. My room here is 118 and K. Paines is 230 where a number of your books are. She closes her house soon. I wish I could see you.

You know the hours here. Dr. Ogden will take me today, that is the only way that seems to keep the water down—so you can see I am better today and worse tomorrow. I have only a few things at K. Paines, but she moves soon—closes house.

Very truly,

H. B. Merrill

*In her final days she was inundated in a sea of reading material which brings to mind Emily Dickenson's verse "There is no Frigate Like a Book to Take Us Lands Away."*

*Harriet Merrill had been continents beyond Dickenson's sequestered realm. Her body was confined but her spirit must have soared when she recalled a heavenly day in sybaritic surroundings of the Rio Botanical Gardens where she had been a guest of the director.*

*Hattie Bell had "Rolled to Rio—Really rolled to Rio, Its wonders to behold, Before she was too old."*

*She died on April 10, 1915, at 20.30. Birge cut the obituary from the paper and saved the notices of her death and funeral. The cause of her death was listed as heart failure. "She became ill several weeks ago, while teaching at the Illinois State University at Urbana, Illinois, and came to Milwaukee. It was believed she was recovering and her death was unexpected." The funeral was held at the First Unitarian Church, Milwaukee, on Tuesday, April 13.*

—by M. L. Hartridge

*Harriet Bell Merrill*
*1863-1915*

# Post Death Correspondence

*In reading letters written to E.A. Birge (State Historical Society of Wisconsin archives) it is interesting to note the number of men in the field of science who apologetically excused their delay in completing assignments by having interruptions from work due to physical disabilities and other set backs. In a poorly typed letter from Frank Smith, March 15, 1915, University of Illinois, he described having seen Harriet Merrill in the hospital when he arrived in Urbana. He thought she appeared better than he expected after hearing that she had difficulty to keep going with her work. "She left the city soon after my visit and I think went to visit relatives in Stevens Point," he wrote.*

*The balance of his letter reveals what other letters in the collection indicated: men were no stronger, resilient or patient than women when confronted with precarious research and living conditions. They certainly found any inequality in pay scales unacceptable. Note that C. Juday, Birge's research assistant, took a lengthy leave of absence due to ill health with no repercussion.*

*The most telling statement in Frank Smith's solicitous letter to Birge was that he had received Birge's interesting paper on the Finger Lakes and thanked him profusely. No mention had been forthcoming on Merrill's input even though she continued to work on the project while gravely ill to edit Birge's paper which was published with credits to him.*

Birge had been in contact with Roger Merrill for some time concerning Harriet Bell. Four letters from Roger to Birge are in the archives at the University of Wisconsin Memorial Library and there may well have been more, each one in response to a letter received from Birge. The first two letters are prior to Merrill's death, the last two in April and May are concerned with the examination and disposal of her writings. Birge obviously decided to keep anything scientific, especially Cladocera.

Although Birge also kept all of Merrill's personal letters to him, from a most comprehensive volume of correspondence, she or her family did not, as we know, keep Birge's letters. However, during her frequent moves in her last years, she mentioned discarding correspondence she no longer had room for.

The first letter to Birge, on February 15, was written eight days after Roger had visited Merrill in the Burnham Hospital in Champaign, at which time he had a long talk with Dr. Newcombe. Merrill's heart had responded to excess strain in the past by dilating. There was no way of telling when from a similar stress it would not be able to dilate. She might live for 20 years, or she might succumb at any time, even from a hearty meal. Roger said she had inherited a weak heart, and he cited instances from their Grandfather Roger Merrill, Colonel W. E. Merrill of Cincinnati, and their own father. "The condition of the latter before his death was exactly the same as my sister's," he concluded.

A letter from Birge on April 7 was opened and answered by Wm. M. Wolff, of Roger's Insurance Agency, when he was in Baltimore. Wolff had seen Merrill on April 6, at which time she was feeling much better than for some time in the past. She asked to be supplied with daily papers, and she was particularly interested in the Milwaukee Free Press because of its favorable attitude toward the University. Merrill obviously was reading, and since Birge was concerned, Wolff assured Birge that she had read all of his letters.

Hotel Belvedere,
Baltimore, MD

March 15, 1915

Dear Dr. Birge:

 Your letter of the 12th has been forwarded to me here. Dr. Ogden called on my sister upon my urging about the middle of the week before last. He found her in worse condition than he had expected. Fluid had naturally accumulated in the chest. There was so much swelling that she could not get her dress together. He immediately ordered her to Columbia Hospital. She felt much distressed, I am sure. She was tapped or something or other done, I think that relieved her. I understood from Dr. Ogden that there did not seem to be the certainty of tuberculosis that he had feared.

 I left Milwaukee on the 7th, Dr. Ogden told me rather unreservedly that I could make any trip with little likelihood of being called back by her condition. Our office writes me that she was sitting up reading on the day after I left, in a chair. Practically nothing would be the matter with her if she would use the least little bit of judgment.

 Sincerely yours,

 Roger Merrill

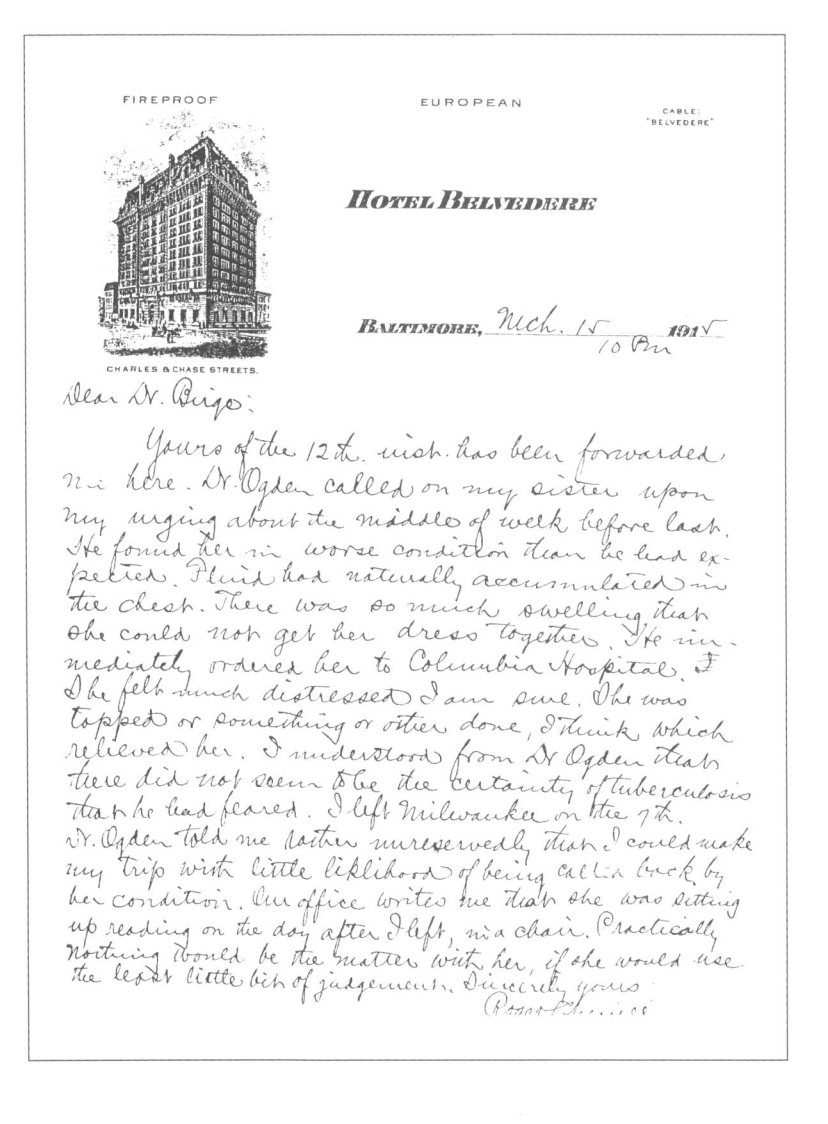

*In a letter of April 24 to Roger, Birge asked questions or presented information on four items, which he numbered. It is not clear what these are about from Roger's response on the 28th. "Matters 3 and 4 referred to negatives and slides." Roger said, "I can do nothing with them except to put them in the attic at the Milwaukee Club. I suggest that they be turned over to you or someone in your department. There is nothing else that I know of Hattie Bell's except at one place, and at the moment I cannot remember the name of the person in charge. Nathan's wife, Marie, will come to Madison in June to look over these belongings with me. At that time we shall probably ask you what can be done with some of them."*

*Roger also reminded Birge, "I have six or eight comparatively small boxes here that I would like to open and go over with you when you are in Milwaukee. I have had the boxes for about a year and they are all in their original condition. There is no reason for any hurry in attending to them. I am only anxious to be at your disposal when it may be convenient for you to go over them."*

*The final letter of May 24 is in response to Birge's of the 22nd, in which he evidently said he could be in Milwaukee that week. Roger was planning to leave for Baltimore at the end of the week, and he said that if he had known Birge could come, he would have stayed over. He expected to be in Madison the second week in June, evidently to look over some effects with his sister-in-law, and he said he would see Birge at this time. He also suggested July as a generally favorable time for everyone to go over Merrill's things.*

*It is well documented that Birge had been closely involved with Merrill when she was alive, and continued to be involved with her, on his initiative, after her death. Eventually the material was looked over. Birge saved her notebooks, and all of her letters from 1902 and 1914-1915 with all of her microscopical preparations important to his discipline.*

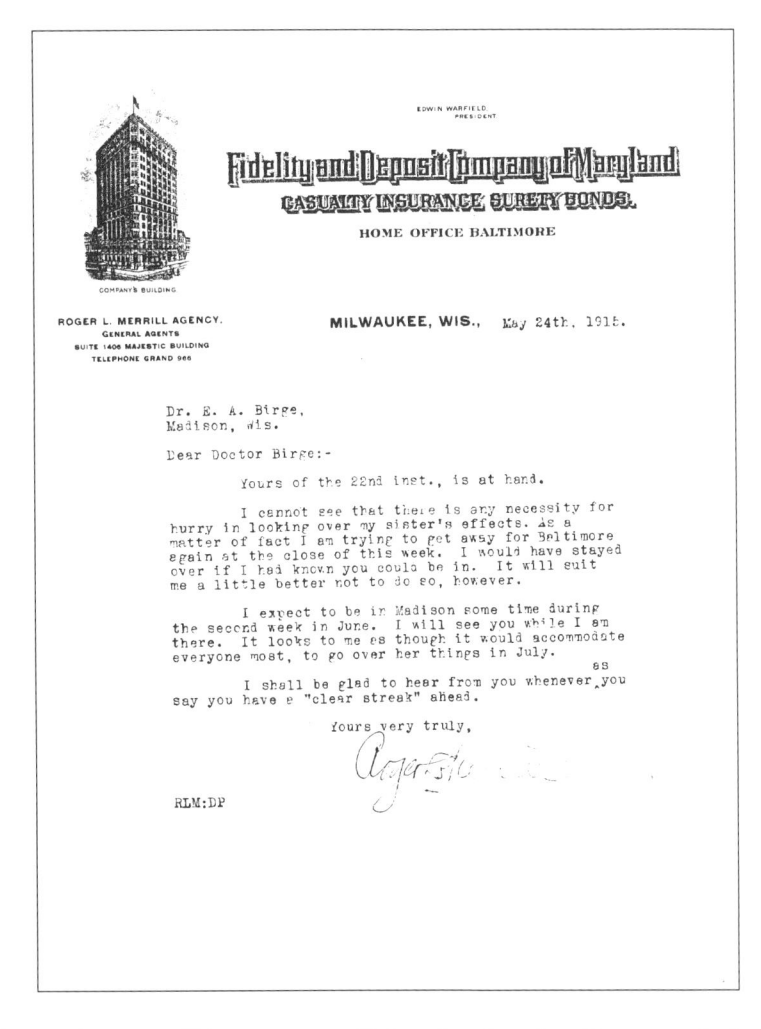

*Because of Harriet Merrill's grave condition, her brother Roger and sister-in-law did not reveal to her the unexpected death of her beloved older brother Nathan, on August 8, 1914, from a ruptured appendix just prior to Harriet's death.*

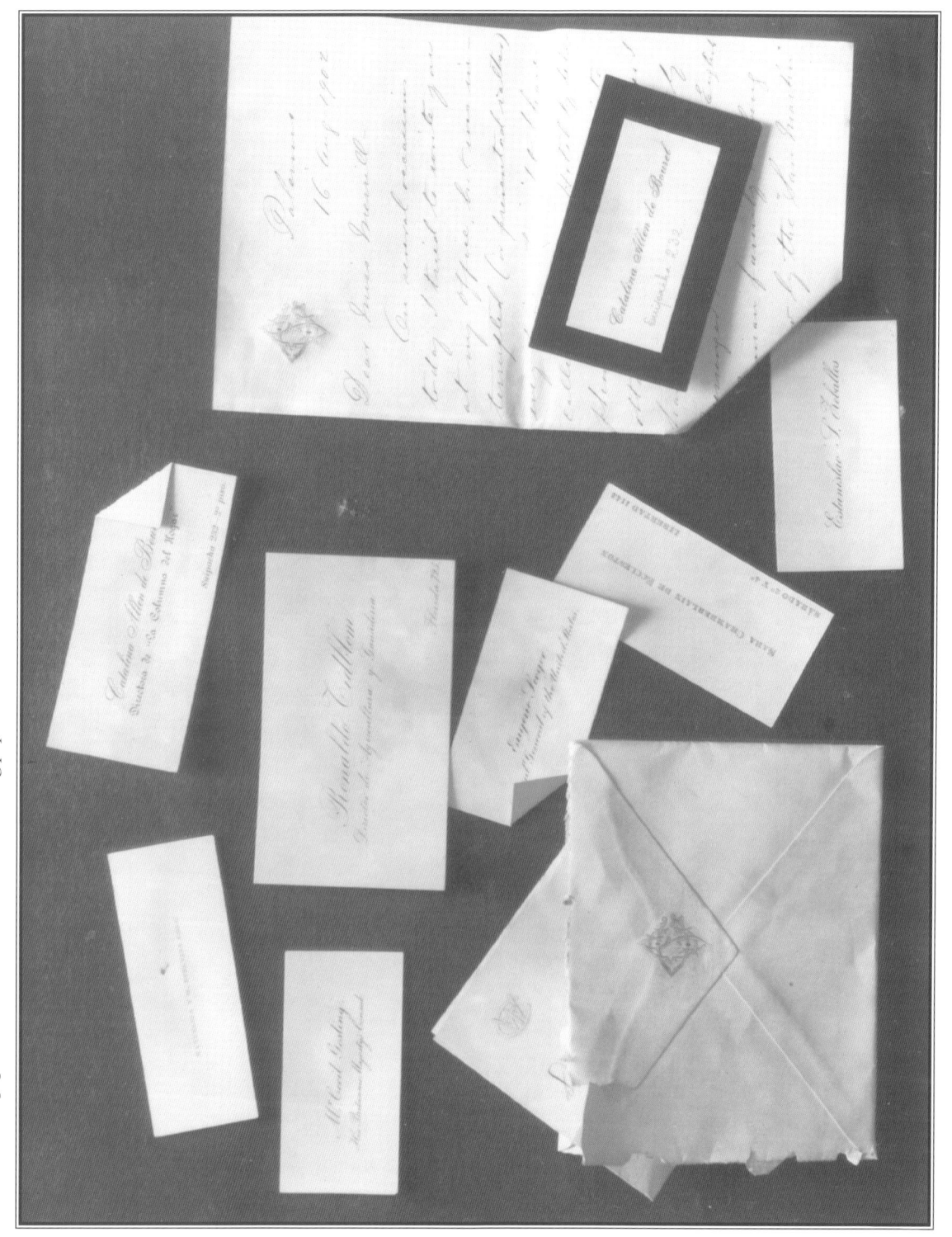

*Cards and note paper used in South America illustrating paper, manners and customs.*

# Part V

# Discovery of the
# E. A. Birge and H. B. Merrill Letters

*Author M. L. Hartridge reveals how the Birge-Merrill letters were discovered and returned to Wisconsin when the mystery of their location was discovered after 72 years.*

Harriet Bell Merrill (Alexander G. Bell descendant) was a persistent, independent woman, afraid of no potential dangers in any country. She met with politicians, educators, scientists, engineers, and native citizens, in South American countries where women unaccompanied by men wouldn't ordinarily have been participants or spectators. Her journeys to South America exposed her to many situations involving substandard lodging, food, and transportation. There were assumptions that her health problems resulted from the exertion and exposure in a foreign environment. This might have been aggravated by an inherited tendency to heart trouble as her brother Roger suggested to Birge.

She knew from what Dr. Ogden predicted that it would be foolhardy to jeopardize her health to leave Oconomowoc for a Ph.D. program at Illinois. It must have been contrary to the sensitivities of the faculty members there for her to attempt to continue in the face of the chronic health concern that had developed. Yet she did, and she suffered the consequences. She died with her boots on, as her forefathers had done before her.

Merrill had an eye for morphological details and a sensitivity to the Cladocera, an organism that would have led to further academic advancement if her health and other circumstances had permitted her to develop her undoubted potential. We have her identifications, in 15 field notebooks, and many of her researched specimens. These are of value to anyone working with the South American fauna. In addition we have an important series of letters, some rather poignant, written to Birge mainly about Merrill's travels, Cladocera, university politics, and health. They were always courteous and respectful, usually rather formal, at least in structure. Birge's replies to them, coming from a major Cladoceran expert in North America, must have kept her fires of hope, burning full flame, during a correspondence of 27 years.

When Merrill returned to the University of Wisconsin as an assistant professor in Zoology, she was hired to work up the results of her South American expeditions at a fee of $600. In addition to her research she was asked to lecture on subjects relative to her associations with educators, economists and governmental personages. The lectures were in the commerce and science departments at the universities of Cornell, Chicago and Wisconsin.

At Wisconsin she was paid $75 per session and her schedule was demanding as she was still working on the scientific findings collected from her expedition. She had kept 15 notebooks from her first trip to South America, with 14 additional field and laboratory notebooks listing; the kind of samples, when and where they were found and their morphological descriptions. My questions on the importance of Merrill's research, began with an inquiry to Dr. Stanley I. Dodson at the University of Wisconsin Zoology Department and Dr. Arthur D. Hasler in the Limnology Center. They suggested that I contact Dr. David Frey, Emeritus Professor of Biology at Indiana University. He had known Birge and was a student at the University of Wisconsin under Chancey Juday. As a noted paleolimnologist and an expert on Cladocera, they assured me Frey was the logical source for related information.

With that lead, I wrote to Dr. Frey. My first letter of July 30, 1986, included background on my Great-aunt Harriet Bell Merrill and names of some of the locations where she had collected Cladocera. I mentioned her monograph on the subject of *Bunops* and asked him about its biological and ecological importance. Along with my letter I enclosed a copy of her article which had been published in the *1893 Transactions* from the Wisconsin Academy of Sciences, Arts and Letters. I was doubtful, that after several decades, Frey would have information on Harriet Merrill's connection with the study. There was nothing in the Birge collection other than a limited number of letters pertaining to the "Birge net" and other letters with a political slant from the period when he was president of the University. The cards and notes from Merrill that had been handed down to me, were of a familial nature, and provided only perfunctory references to her association with Birge.

Dr. Dodson also had informed Frey of my search for more information on Merrill and Frey then puzzled over the series of personal letters that had been sent to him along with a number of field and laboratory notebooks from Dr. E. S. Deevey's laboratory at the University of Florida. Until receiving my request, he had not known of Merrill's exact connection as assistant to Birge. As he remarked, "the coincidence of receiving a letter from Merrillyn Hartridge inquiring about the Birge connection with Merrill was almost staggering." Frey stated that the Birge-Merrill letters revealed a deep relationship between the two for over 27 years. He also learned from the notebooks that Merrill had two collecting expeditions to South America and had worked over the material on her return to Wisconsin. "The notebooks have enabled the decoding of many specimens which had for decades been indicated only by number in the Birge collection," he said.

Frey indicated that some letters might have gone astray from the time of Merrill's death in 1915, but Birge obviously thought her notes, books and letters important enough to have kept all of them for reference to her work. Frey had received the material from one of Dr. Deevey's assistants who recognized that much of it was research on Cladocera and Deevey, now retired, was about to turn all of the papers over to the Yale Library. Learning of Frey's sudden resurgent interest (1986), to use them for his research, Mike Binford, Deevey's student assistant, had them sent to Bloomington, Indiana. In writing also to Dr. Deevey, I hoped to learn more on the mystery of the transference of the Birge-Merrill letters. His reply was that Arthur Hasler had probably sent the material along with Juday's notes which had been designated to go to John L. Brooks, a noted expert in Cladocera study, at Yale University. There was also the possibility that some of the records might have been diverted at the time of distribution of the papers from the Juday estate. Mrs. Juday was displeased with Birge's neglect to compensate her husband with retirement pay for his work in the Geological and Natural History Survey and Birge had not consulted them on the disposal of the Juday papers. As a result, she sent her husband's books to the Academy of Natural Sciences in Philadelphia. *[See chapter notes, page 192: Chancey Juday—A Birge Associate]* All of this was brought to my attention as an explanation of how research material can be diverted from one laboratory to another for future application of study or in some cases, mislaid on some dusty shelf and lost to science.

In his letter of August 1, 1986, Dr. Frey, described the species *scutifrons* in the genus *Bunops* written by Merrill and published in the 1893 *Transactions* from Wisconsin Academy of Sciences, Arts and Letters as, "a new genus in 1893. According to information gained from Merrill's notebooks, nothing had been known about the diversity of the species in South America prior to her findings." Frey himself said that he had found them only a couple of times. From Argentina where he traveled in early 1989, Frey wrote that, "Dr. Stillman Wright who received his Ph.D. from Juday in 1928, had access to Merrill's collections from Brazil and named a species of *Diaptomus* from that country after her…the *Merrilli*." "The importance of these forms of life," he explained, "is their relationship to the ecology of the earth's bodies of water. They occur in all fresh waters on all continents except Antarctica. Their most important function is to convert bacteria, algae and detritus material that can be utilized by larger invertebrates. The balance of these microscopic protozoans in our fresh water systems, determines the life or death of the major bodies of water in the earth's environment." It is understandable how native South Americans referred to Merrill as "the brave little lady who hunts the unseen."

David Frey described her research on the "unseen" as unflagging. "Her grasp of the most minute differences of species and their populations is even today, exciting to behold. Her identification of Chydoridae and Macrothricidae, exceeds 82 taxa and the work involved in collecting and examining that number of samples and making detailed notes, was tremendous. One can only admire the initiative and determination that motivated such a monumental task."

I am grateful to Dr. Frey for adding a scientific dimension, vital to my biography based on my letters and the Birge-Merrill letters. If I had not made contact with him when I did, I was told that the connection with Merrill as assistant to Birge might be lost or known only through the Yale Library. When David Frey asked my opinion on where the letters should be placed, I suggested the State Historical Society of Wisconsin. He agreed that it was more logical than at Cornell or Chicago and was eager to put the letters into my hands knowing how anxious I was to examine them.

---

**INDIANA UNIVERSITY**
*Department of Biology*
JORDAN HALL 138
BLOOMINGTON, INDIANA 47401

TEL. NO. 812—337-

19 June 1987

Dear Merrilyn:

Many thanks for meeting with me on such short notice and at such an awkward time. I enjoyed the meeting and hope this is just the first of many.

I have enclosed xerox copies of the notes I made on her letters to Birge. At first I was concerned only with the chydorids, but rapidly got into other interests and her close contacts with Birge. She is a most interesting person.

Sincerely,
David Frey

When Frey came to Madison with the Merrill letters, in June of 1987 it was an occasion that marked the beginning of several years of correspondence between us. I had learned from him the importance of the contribution my great aunt had made in her field of science and he "met" Miss Merrill through my biographical portrait. Frey described her as most attractive. He wondered if I had any clue as to whether she and Birge might have had more than a professional interest in each other. My family referred to her as pretty and sweet, a soft-spoken lady with enormous grey blue eyes. I had to admit that her pursuits did show a certain tenacity and spunk, but Frey and I agreed that anyone who had accomplished all that she did, would have had no time for romantic liaisons and marriage. In her time, it would have been impossible.

*Paleolimnologist Dr. David Frey and author M. L. Hartridge.*

Upon his return to Indiana, Dr. Frey sent a handwritten thank you note (6/19/87), with wishes that we might continue to compare information found, regarding my Aunt Hattie. On July 8, 1987, Dr. Frey wrote that Merrill's life was an interesting story and felt, as did persons in the Limnology Department, that she had been neglected too long. On July 16, 1987, Frey asked again, that I send my biography of Merrill. He wrote back that he had found my compilation of Merrill genealogy, family records and history of the times most fascinating reading. On August 18, 1987, he reiterated in a letter, the conclusion he had made previously that, "Merrill became a well respected scientist by the time she completed research in South America." He lamented the unfortunate circumstances that led to her deteriorating health after she returned to the states, and conjectured as I had, that certain conditions she faced in the tropics might have been the cause. "She had become a real pioneer in the study," he said.

# The First Inquiry to Dr. Frey

July 30, 1986

Dr. David G. Frey
Dept. of Zoology
Indiana University

Dear Dr. Frey:

During my research on H. B. Merrill who was an assistant professor of Zoology at the University of Wisconsin in the late 1800s, I find that she is accredited with isolating certain phylum Arthropoda. Not being familiar with Cladoceran genus, I asked Professor Stanley Dodson, U.W. Madison-Zoology department, if he could help me define the taxonomy and the importance of Merrill's findings. He suggested it is probably Macrothricidae but referred me to you as the best source in the country to recognize the name of the genus found by professor Merrill. Was the *Bunops scutifrons* considered rare in 1892?

Miss Merrill had collected nonpelagic material from various fresh water sources in Europe and found little if any differentiation in the anatomy of the microorganisms. From 1902-03 and 1907-09, she traveled extensively throughout South America to collect species of crustaceans which were acquisitioned by the UW and Milwaukee Museum of Natural History. She continued to pursue the aquatic microorganisms in South America. One of my questions is: what was the ecological or biological importance of her find? Was it simply scarce? While in South America, Merrill questioned prominent doctors of medicine who were developing antitoxins and beginning advances on attenuated viruses.

During her lifetime, Assistant Professor of Zoology, Merrill, was given scant recognition for her devotion to the sciences and tireless dedication to her teaching. She was a lecturer as a Fellow at Cornell and Chicago Universities and continued teaching until her death at age 52 from myocarditis, later thought to have been contracted during her South American expedition. Her findings were published in various scientific journals which included meticulous drawings of microorganisms isolated by Merrill.

I would greatly appreciate any light you can shed on the above questions. I want to incorporate them into a biography about her.

Along with scientific findings, it is interesting to note other parallels in viewpoints on university and national politics as expressed by Merrill to Birge as well as by other colleagues. Her assessments of faculty personalities though mild compared to her counterparts, makes for fascinating reading and becomes a personal glimpse into her era. Although she was considered a didactic individual who pushed herself beyond her physical endurance, her letters prior to her illness, reveal a zest for life and an eagerness to live it as one, grand "expedition." She came from New England stock (1630s) and was interested in the sciences at an early age. Her purpose in continuing to study was to eventually head a woman's college. As I mentioned previously, she was head of sciences at Milwaukee-Downer College at one time. Unfortunately, when these other offers came her way, she was too ill to pursue them.

It is my feeling at this point, that Merrill had a "blind devotion" toward Birge. It was professional admiration, manifest in her steadfast support of his work. You might recall that she was critical of colleagues who misquoted him. I read a letter to Birge from one of his collectors who had devised a new kind of net. Then I recalled that Birge had sent Merrill the "new" net with which she was pleased and was certain that he could find a market for them. His correspondence did not indicate that he was generous with credit to colleagues. Birge must have encouraged Merrill to some extent, for her to have responded with as much personal information as she did. He undoubtedly felt some comfort in her friendship. Her arduous journey to South America was encouraged by him. They corresponded from wherever she was. The inconsistencies in the life of a woman at that time was that he…could be married. She would have been dismissed from her life's work if she had married, even if she had no children. I still am interested in what they were trying to establish in their field of science. I found nothing notable on Birge. This could be an ongoing quest, leading to research on papers of the associates connected with Merrill.

I will welcome anything further you can share on the subject. I am enclosing an amount to defray some of the cost of xeroxing and postage. Please let me know if it is not adequate.

Sincerely,

Merrillyn Hartridge

*Within a day of writing to Dr. Frey, I received his reply. It was dated August 1, 1986, and began with an incredulous answer to my inquiry.*

INDIANA UNIVERSITY | DEPARTMENT OF BIOLOGY
Jordan Hall 138
Bloomington, Indiana 47405
(812) 335-

1 August 1986

Merrillyn Hartridge
Madison, WI 53716

Dear Ms Hartridge:

Your letter, just received, almost made me gasp, because just this morning I had been reading a series of letters written by Merrill to Birge in the period 1914-1915 up to and including her death. At that time she was at the University of Illinois as a student of Zeleny, although her addresses during this period fluctuated from Oconomowoc to Urbana to Milwaukee. She wrote Birge frequently then, sometimes as often as every other day. Unfortunately, I have only her letters to Birge, not his replies to her, which he felt important enough to have kept.

I cannot answer all your questions about it. The species *scutifrons* in the genus *Bunops* were both described by Birge in 1893. Daday of Hungary had previously described a species from Hungary as *Macrothrix serricaudatus*, which Birge decided also belonged to his new genus. Thus, there are two species in the genus *Bunops* now, one species in Eurasia and the other in North America. The North American species has not been collected very often, possibly because it occurs in habitats that limnologists usually do not pay any attention to, and Merrill assiduously collected. Besides the records from northern Wisconsin and the Madison region, it is also known from Minnesota but where else I cannot say. Possibly we have collected it in the Maritime provinces and in Maine. If you wish I shall search our records to see where we have it from.

Birge's description is minimal. He states that details can be obtained from Merrill's paper, which is in the same volume of the Transactions as his. Merrill's paper is excellent. She has detailed observations on the external and internal anatomy, but of greater significance she has arranged all the genera of the family Macrothricidae according to differences in number of setae on the antennae, structure of the postabdomen and shape of the antennules.

This is a long letter, which probably doesn't give you much of what you want about Merrill. Both of us are interested in her, you for very obvious reasons and me for scientific reasons. Her relationship to Birge seems completely fascinating, yet Sellery does not mention her in his book. Have you developed an insight into this relationship?

My home was in Wisconsin at Hartford. Just last weekend we visited there to see my brother and at Appleton to see my sister. If I had been aware of your interests then we could have come through Madison on our way back to Bloomington. The distance is not great, and if the occasion demands, I might well drive up that way again.

[Dr. Frey did shortly thereafter.]

Sincerely,

David G. Frey

P.S. Please thank Stan Dodson for suggesting me to you. The association will be valuable to both of us.

*Frey stated that he too wanted to write about her and requested details on her life that he did not have. In subsequent letters, he asked me to make corrections if needed in the additions to his paper. He kept in touch while continuing his research, even giving me a forwarding address to the Department of Botany when he was at the University of Adlaide, "Just in case anything more comes up." In his paper on Merrill, Frey stated, "The knowledge I have on Harriet Bell Merrill has come about only because of the research done by Merrillyn Hartridge on the history of the former University of Wisconsin scientist. I knew nothing of her personal life and this connection has heightened my understanding of this remarkable woman."*

*In between his travels from one continent to another, his correspondence detailed his theories on the origins and genetic modifications of Cladocera. In 1990, Dr. Hartridge and I had the pleasure of visiting Indiana University and Dr. Frey took us to his laboratory to view some of his work and a close-up of the Bunops. David G. Frey—who helped resurrect the memory of a woman who had preceded him in his life's work seventy-five years ago, died in April of 1992.*

August 12, 1986

Dr. David Frey
Department of Biology
Indiana University

Dear. Dr. Frey:

I was delighted and somewhat astounded to read what I perceive to be rather personal notes from Merrill to Birge. I promptly studied them all and have spent the last two days in the Birge archives. I found that most of the correspondence directed to him was related to administrative or political matters. He was, it appears, quite ambitious to advance his position in the university and Merrill was obviously one of his staunchest supporters for office of president.

The only correspondence filed in the S.H.S.W. archives on Birge are those addressed to him. There may be others at Memorial Library, and I will get to them when the librarian returns from vacation. The bulk of the letters seem to be from Cornell and Chicago Universities and a source in New Orleans. Of those, I have selected a few at random to send to you. Although the volume was not as prodigious as I had expected, Birge seems to have accomplished a large part of his research through a constant correspondence with several collectors in his field. They all seemed to be searching for specific keys. You are probably familiar with the 1912 book which Merrill referred to, *Description of Recently Discovered Cladocera from New England* by A. A. Doolittle.

Sincerely

Merrillyn Hartridge


*Harriet Bell Merrill worked with Henry Baldwin Ward; and I found a considerable amount of correspondence to Birge from him, echoing some of the theories that Merrill proposed. I did not come across her name in the papers I examined however. As was often the case, female assistants were expected to serve, not take charge. With all of the assistance Birge received, his name often topped the list of credits on published works. For example, Merrill had contributed as much or more as her male colleagues to the studies on Inland Lakes. (See page 145 for a letter from Merrill to Birge dated November 25, 1914.)*

**INDIANA UNIVERSITY** | DEPARTMENT OF BIOLOGY
Jordan Hall 138
Bloomington, Indiana 47405
(812) 335-

*Response of 10/10/86 from Frey to Mrs. Hartridge*

Dear Merrillyn:

As for myself, I studied with C. Juday at Wisconsin during the period 1932-40, and got to know Birge during this period. At that time I was working with fish (carp) in the Madison lakes and had almost no interest in Cladocera. The interest developed when, on doing a paleolimnology study of a lake in Austria in 1954, I found these many membranous bits of chitin that later proved to be parts of the exoskeleton of Cladocera, principally one family. Since then I have been working mostly on Cladocera, and now I am doing only that. Other scientists for the past 100 years or so have been assuming that the Cladocera are cosmopolitan, and for this reason they have not looked at them closely enough. We have been finding instead that species having the same names on different continents, when studied closely, are found to be different species. They definitely are not cosmopolitan, and their present distribution cannot be explained by the passive transport of resting eggs. A potential alternate explanation is that the main features of the present distribution were generated millions of years ago by the progressive splitting off of daughter continents from Gondwanaland. That is why I am going to Australia and New Zealand in three weeks, and that is also why I shall be returning to southern South America in a couple years to obtain some definitive collections from there as well.

I think I told you that I enjoyed reading the several pages from your Merrill's journal, which I presume is now published. Could you please provide a reference for me as I should like to order the book.

Many thanks again for all your help. I hope that both of us succeed in finding more material pertinent to our respective subjects regarding Merrill.

Sincerely,

David G. Frey

*Response of 3/17/89 from Frey to Mrs. Hartridge*

The Cladocera, especially members of the families Chydoridae and Bosminidae, leave their exoskeletal remains in the sediments, from which the species that produced them can be identified positively. These animals, like all related forms (insects, spiders, lobsters—all invertebrates with jointed legs and their skeletons on the outside), can grow in size only by periodically shedding their external skeleton and expanding rapidly before the new skeleton has a chance to harden. The old exoskeleton or exuviae is what preserves in the sediments. These remains are known back into the Tertiary and even the Cretaceous. They are part of my basis for believing that I can find species in the southern continents that are related to the breakup of Gondwanaland, beginning about 150 million years ago. The remains in the sediments are good because they give us documentation as to what species existed in a lake in past time, and what their relative importance was, one species to another. Species of these animals are eliminated from a habitat by changes in conditions that the animals cannot tolerate. Thus we can study the changes over time in sediments and get some idea of the changing conditions in the lake, knowing what conditions each species indicates.

*Frey continued:*

Birge had almost no students, and certainly no Ph.D. students. Thus it is true to say that H.B. Merrill was the only one (of Birge's students) who travelled so far afield. But Stillman Wright, who received his Ph.D. from Juday in 1928, is another. He lived in Brazil for a number of years where he collected copepods—another group of small freshwater organisms. I learned during my trip to Argentina that he had access to the Merrill collections from Brazil because he described eight new species of *Diaptomus* from that country, one of which he named *Merrilli* after her.

There is so much about Cladocera that I could go on for pages and possibly not get into material of real concern to you. If you have any specific questions, I shall be glad to try to answer them.

The Allen "mess" refers to a survey of the University made by Allen in the early teens. It is described well in the book on Birge by Sellery published in 1956 by the U of W Press.

I hope that you are well and getting back to more satisfying activities.

Sincerely,

David G. Frey

INDIANA UNIVERSITY | DEPARTMENT OF BIOLOGY
Jordan Hall 138
Bloomington, Indiana 47405
(812) 335-

June 19, 1987

Dear Merrillyn:

Many thanks for meeting with me on such short notice and at such an awkward time. I enjoyed the meeting and hope this is just the first of many.

I have enclosed xerox copies of the notes I made on her letters to Birge. At first I was concerned only with the *chydorids* but rapidly got into other interests and her close contacts with Birge. She is a most interesting person.

Sincerely,

David Frey


*Birge is not mentioned in Merrill's notebooks, and there is no suggestion that he was involved in her decisions. She obviously functioned on her own. Nothing was known about the diversity of the species in South America until the discovery of H. B. Merrill's notebooks.*

*There are 15 notebooks in all, one of which is from the 1902-03 excursion. The other 14 are field and laboratory notebooks (1907-09), in which she has the individual samples listed with details of place, date, and general description of each station followed by a list of the taxa recovered, with notes on their abundance, morphological details, and other aspects. She was well aware of males and ephippial females, looked for them especially, recorded their occurrence, and described their seasonal distribution.*

*Her notes are full of comments about size and number of teeth on the labrum and about the number of teeth at the posterior-ventral angle of the shell. Sometimes a tooth is completely lacking here, and sometimes there is just a single stout bristle, which cannot be considered a tooth. What all this indicates is that there is a considerable number of species of this genus in South America as well as in the extreme southern United States.*

Table 1. Approximate summary of samples collected in South America by Harriet Bell Merrill and the number of taxa she identified in each group.

| Year | Dates | Country | Place | Waterbodies | No. of samples | Approx. no. of species |
|---|---|---|---|---|---|---|
| 1902 | ca. 19 July | Brazil | Pernambuco | | 71 | 16 |
| | 20 July | | Bahia | inland waterbodies | | |
| | 21-31 July | | Rio de Janeiro, Petropolis | | | |
| | 1-7 Aug. | | São Paulo | Campo Grande, Horte Botanica, Ponte Grande, ditch, weedy ponds, rivers | | |
| | 8-28 Aug. | Argentina | Buenos Aires | Cordoba St. Pond, Tigre, Zoological Gardens, Darsena | | |
| | 28 Aug.-25 Sept. | Paraguay | Asuncion | San Bernardino Lake, plus nearby ponds and ditches, Pasadas | | |
| 1907 | 17 Aug. to at least 20 Oct. | Brazil | Marajo | Cachoeirna, Lake Arang, Diamentina pond | 152 | 63 |
| | December | | Pará | Botanic Garden, Bosque Park, Igarape, Peixe Boi, Val de Cams | | |
| 1908 | Jan.-8 Mar. | Brazil | Manáos | Flores, Igarape Fletcher, cahoerina, Lake Santa Maria, São Antonio, Paraense, Manicore, Cathedral Parl, Prosa trinta seis tank | 79 | 40 |
| | 25 Mar.-4 May | | Rio Madeira | Lake São João, igarape, Madena River, Humaytá, Fio Meza | 50 | 46 |
| | 9 May-5 Aug. | | near Calama and Rio Machado | igarape, tanks, puddles, Calama Island, small lake at Mannins, Lago Verde, South Pond, cachoerinas | 77 | 41 |
| | 10 May-16 Sept. | | Calama | South Pond | 91 | 47 |
| | 12-22 Oct. | | Amazon River near Santarem | lake or igarape across Tapajos, Lago Grande de Monte Alto | 55 | 40 |
| | 12-20 Nov. | | Rio Aramá | Villa Dieo | 22 | ? |
| | 11 Dec. | Venezuela | Orinoco | near Ciudad Baliras (?) | 7 | 21 |
| 1909 | 4-9 Feb. | Trinidad | | Pitch Lake, Sange Grande | 21 | 18 |
| | 17 Feb.-9 Mar. | British Guiana | Georgetown | Essequibo River, Indian Camp at Ituri Cr., Savannah Lake, Lake Co, canals near Longdon Park, Barrock St. canal, Parade St. canal, canals near Christ Church, Irving St., Port Limon, Lamaha canal | 57 | 51 |
| | 16 Apr. | Curaçao | | | 3 | 6 |
| | 25 Apr. | Venezuela | | Los Estanislaus, Magdalena River | 3 | 6 |
| | 8-17 May | Panama | | Lagoon across Chagres River, lagoon opposite Culebra cut, Savannah-Old Panama Road, Paraiso near Trask, Mt. Jennings Laboratory | 14 | 20 |

[1] In addition from British Guiana there are also 18 samples from the Gimbel Collection of 15 Sept.-3 Oct. 1910, and 2 from the Owen Collection in 1901.

Table 2. List of taxa of Cladocera reported by H.B. Merrill from the samples she collected in South America.

SIDIDAE

    Diaphanosoma sarsi
        "    n. sp.
    Latonopsi fasciculata
        "    orientalis
    Pseudosida bidentata
        "    tridentata
        "    ramosa

DAPHNIIDAE

    Ceriodaphnia rigaudi
        "    cornuta
        "    reticulata
        "    n. sp.
    Scapholeberis mucronata
    Simocephalus serrulatus

MOINIDAE

    Moina micrura
        "    with eye spot
        "    without eye spot
    Moinodaphnia macleayi

MACROTHRICIDAE

    Ilyocryptus
    Grimaldina
    Iheringula paulensis
    Guernella
    Streblocerus
    Macrothrix elegans
        "    laticornis
        "    squamosa
        "    small
        "    large

BOSMINIDAE

    Bosmina hagmani
    Bosminopsis

CHYDORIDAE

    Eurycercus
    Acroperus harpae
    Camptocercus
    Alonopsis
    Pseudalona longirostris } (= Kurzia)
        "    latissima

CHYDORIDAE (cont)

    Odontalona longicaudis } (= Oxyurella)
        "    tenuicaudis
    Graptoleberis
    Alona monacantha
        "    glabra
        "    cambouei
        "    rectangula
        "    novaezealandiae
        "    pulchra
        "    costata
        "    guttata
        "    intermedia
        "    verrucosa
        2 n. sp.
    Leydigia acanthocercoides
        n. sp.
    Leydigiopsis curvirostris
        "    megalops
        "    n. sp.
    Dadaya
    Alonella nana
        "    clathratula
        "    dadayi - 2 spp.? } (= Disparalona)
        "    rostrata
        "    diaphana
        "    karua     (= Alona)
        "    dentifera
        "    globulosa or sculpta (= Notoalona)
        "    nitidula
        "    n. sp.
    Dunhevedia odontoplax
    Pleuroxus hamulatus ( = Alonella)
        "    scopuliferus
        "    hastirostris
        "    n. sp. -- long labrum
    Chydorus barroisi (group)
        "    poppei     (= Ephemeroporus)
        "    hybridus
        "    globosus (= Pseudochydorus)
        "    pubescens
        "    eurynotus
        "    ovalis
        "    faviformis
        "    sphaericus
        "    2 n. sp.

Table 1. Approximate summary of samples collected in South America by Harriet Bell Merrill and the number of taxa she identified in each group.

| Year | Dates | Country | Place | Waterbodies | No. of Samples | Approx. no. of species |
|---|---|---|---|---|---|---|
| 1902 | ca. 19 July | Brazil | Pernambuco | | | |
| | 20 July | | Bahia | inland waterbodies | 71 | 16 |
| | 21-31 July | | Rio de Janeiro, Petropolis | | | |
| | 1-7 Aug. | | São Paulo | Campo Grande, Horte Botanica, Ponte Grande, ditch, weedy ponds, river | | |
| | 8-28 Aug. | Argentina | Buenos Aires | Cordoba St. Pond, Tigre, Zoological Gardens, Darsena | | |
| | 28 Aug-25 Sept. | Paraguay | Asuncion | San Bernardino Lake, plus Nearby ponds and ditches, Pasadas | | |
| 1907 | 17 Aug. to at least 20 Oct. | Brazil | Marajo | Cachoeirinha, Lake Arang, Diamentina pond | 152 | 63 |
| | December | | Pará | Botanic Garden, Bosque Park, igarape, Peixe Boi Stream, Val de Cams | | |
| 1908 | Jan.-8 Mar. | Brazil | Manaos | Flores, Igarape Fletcher, cachoeirina, Lake Santa Maria, São Antonio, Paraense, Manicore, Cathedral Park, Prosa trinta seis tank | 79 | 40 |
| | 25 Mar.-4 May | | Rio Maderia | Lake São João, igarape, Madena River, Humayta, Fio Meza | 50 | 46 |

| Year | Dates | Country | Place | Waterbodies | No. of Samples | Approx. no. of species |
|---|---|---|---|---|---|---|
| 1908 | 9 May–5 Aug. | | near Calama and Rio Machado | igarape, tanks, puddles, Calama Island, small lake at Mannins, Lago Verde, South Pond, cachoerinas | 77 | 41 |
| | 10 May–16 Sept. | | Calama | South Pond | 91 | 47 |
| | 12–22 Oct. | | Amazon River near Santarem | lake or igarape across Tapajos, Lago Grande de Monte Alto | 55 | 40 |
| | 12–20 Nov. | | Rio Aramá | Villa Dieo | 22 | ? |
| | 11 Dec. | Venezuela | Orinoco | near Ciudad Baliras (?) | 7 | 21 |
| 1909 | 4–9 Feb. | Trinidad | | Pitch Lake, Sange Grande | 21 | 18 |
| | 17 Feb.–9 Mar. | British Guiana | Georgetown | Essequibo River, Indian Camp at Ituri Cr., Savannah Lake, Lake Co, canals near Longdon Park, Barrock St. canal, Parade St. canal, canals near Christ Church, Irving St., Port Limon, Lamaha canal | 57[1] | 51 |
| | 16 Apr. | Curaçao | | | 3 | 6 |
| | 25 Apr. | Venezuela | | Los Estanislaus, Magdalena River | 3 | 6 |
| | 8–17 May | Panama | | Lagoon across Chagres River, lagoon opposite Culebra cut, Savannah-Old Panama Road, Paraiso near Trask, Mr. Jennings Laboratory | 14 | 20 |

[1] In addition from British Guiana there are also 18 samples from the Gimbel Collection of 15 Sept.–3 Oct. 1910, and 2 from the Owen Collection in 1901.

### COLLECTIONS OF THE PUBLIC MUSEUM
#### DEPARTMENT OF ZOOLOGY.—REPTILES & BATRACHIANS.

| Entry | Current Number | Original Number | Room | Case | Drawer | No. of Spec. | Name | Sex | Measurement | When Coll. |
|---|---|---|---|---|---|---|---|---|---|---|
| Dec 26 | 1778 | 336 | | | | 0 | Alligator mississippiensis (Dundi) | | | Rock? |

### COLLECTIONS OF THE PUBLIC MUSEUM
#### DEPARTMENT OF ZOOLOGY.—MOLLUSKS.

### COLLECTIONS OF THE PUBLIC MUSEUM
#### DEPARTMENT OF INSECTS

### 1915 COLLECTIONS OF THE PUBLIC MUSEUM
#### DEPARTMENT OF ANTHROPOLOGY : DIVISION OF ETHNOLOGY

### 1915 ADDITIONS TO COLLECTIONS OF THE PUBLIC
#### DEPARTMENT OF BOTANY

| Date of Entry | Collector's No. | Current No. | Accession No. | No. of Spec. | Name | Group | Where Collected |
|---|---|---|---|---|---|---|---|
| May 17 | 8 | 45276 | Part of 5392 | 1 | "Cupiuba" | Wood | Calama, Rio Madei |
| " | 9 | 45277 | " | 1 | "Aquariquara" | " | " |
| " | 10 | 45278 | " | 1 | "Lauro-amarello" | " | " |
| " | 11 | 45279 | " | 1 | "Pereiora ou Casca-preciosa" | " | " |
| " | 12 | 45280 | " | 1 | "Louro-coqueiro" | " | " |
| " | 13 | 45281 | " | 1 | "Cedro vermelho" | " | " |
| " | 14 | 45282 | " | 1 | "Cedro roxo" | " | " |
| " | 15 | 45283 | " | 1 | "Bouro-rosa" | " | " |
| " | 16 | 45284 | " | 1 | "Jacareuba" | " | " |
| " | 17 | 45285 | " | 1 | "Marupa" | " | " |
| " | 18 | 45286 | " | 1 | "Piranheira" | " | " |
| " | 19 | 45287 | " | 1 | "Pequirana" | " | " |

*Acquisitions from H.B. Merrill's scientific collection. Courtesy of Carter Lupton, curator, Milwaukee Public Museum.*

# Chapter Notes
## Researched and compiled by M. L. Hartridge

### 1. Reflections of Birge by Colleagues *(from page 2)*

Dr. Birge was a teacher, scholar, administrator and scientist. The tributes which were paid to him in 1925 by faculty colleagues, in 1940 by colleagues and other friends at the Symposium on Hydrobiology, and in June 1950, at Commencement by President Fred, Vice President of the Regents A. Matt Werner, Dean Ingraham, and others at the dedication of the Biology Building as Birge Hall, show the high esteem in which he was held by those who knew him well. And quite properly none of these tributes, all of them except those of the first group published by the University, makes mention of certain peculiarities which must be considered if we are to understand him as fully as possible. For there were, let us admit it, spots upon the sun.

Dr. Birge had little if any of the "small talk" or light conversation which is so useful at dinner parties and on other social occasions, and even in office meetings with students and with colleagues. That observation largely explains the side of Dr. Birge which we are considering, but it does not quite explain the tartness or "bite" of remarks which have been illustrated in earlier pages of this memoir.

The reader may recall the story related in Chapter III of H. J. Thorkelson's visit to Dr. Birge's office to get excused for absence from class, how Dr. Birge had and kept his head bent over a microscope, barked "all right" to Thorkelson's plea and left him to discover to his embarrassment that the interview was over. That story, *mutatis mutandis*, could be repeated by quite a number of students. They waited in vain for words of dismissal; they got out of the office as best they could. And the same could be said, unfortunately, of the experience of professors who were occasionally put out of countenance by similar treatment.

Is the explanation of this trait to be found partly in a lack of personal warmth on the part of Dr. Birge? In 1899 President Adams—a great admirer of the Dean—wrote at length as we know, to the authorities of the University of Iowa who were looking for a new president and had asked about Dr. Birge. On the point we are considering Mr. Adams wrote: "By some his nature is thought to be cold and unsympathetic. Perhaps this would be the judgment even of some of his friends; but is, I think, rather to be regarded as an indication that he keeps his emotions in reserve. From these characteristics you will naturally infer that by his colleagues and friends he is not so much beloved as admired and respected. This perhaps would describe his relation to students as well as to his colleagues and the public at large."

There were some people on the staff who would add to Dr. Birge's lack of personal warmth, a streak of positive harshness of something approaching ruthlessness. Certain it is that he found it harder to praise than to dispraise. His occasional denunciation of opponents in stiff faculty contests, as the writer can testify from personal experience at his hands, was fairly sweeping—and not entirely palatable. But such a display of temper—or indignation—was a rare event. President Adams was right: "*[Dr. Birge]* keeps his emotions in rather remarkable reserve."

—George Clark Sellery, *Edward Asahel Birge a Memoir* with an "Appraisal of Birge the Limnologist" by C. H. Mortimer, University of Wisconsin Press, Madison, Wisconsin, 1956, pp. 146- 147.

Charles Kendall Adams had said that Birge was thought by some people to be cold and unsympathetic. Many could testify that he had a sharp tongue and a sharper pen, that his wit was often mordant, sometimes brutal. When the business manager of the *Octopus*, the student humor magazine, wrote to him apologizing for an advertisement which had appeared in the *Cardinal*, Birge accepted the apology but remarked that "it had never occurred to me that it was notably below the standards of the *Octopus* as regards either taste or wit;" he suggested changing the name of the *Octopus* to the *Tea Hounds' Review* since its chief aim seemed to be "to give the university 'tea hound' a place in which he may publicly show off his uncontrollable amorous propensities, after the manner common to his kind." He reserved his most pungent comments for the makers and circulators of questionnaires. To a young instructor at the University of Michigan who was collecting information on "the modern American university as a social organization," Birge wrote saying that while he never found any pleasure in questionnaires, "I am bound to say that yours seems to be a peculiarly obnoxious one." Birge interpreted a question on socialism as a trap for the unwary. He himself refused to be trapped, but he charged the author with being naive and remarked, "Indeed one can hardly help wondering how a person with so great naivete has reached the position of instructor in the University of Michigan." To another information seeker, he wrote, "I wonder how much you can get out of a questionnaire of this sort? I must own that it looks to me something like the survival of the Laputan idea of extracting sunshine from cucumbers. I am obliged to confess that I am not that kind of cucumber. However, let me say on the other side that while I am receiving questionnaires in practically every mail, yours is in a new form and I think you may be congratulated on adding a new terror to human life—at least so far as human life is shared by college presidents." Frederick B. Robinson, secretary-treasurer of the Association of Urban Universities, wrote to President Van Hise, reporting that the University's dues had not been paid. Birge replied that the University of Wisconsin was not a member and was not in a position to contribute much to the discussion of problems of urban universities. "It is, of course, regrettable that institutions should thus fall into classes, but the absence of close connection between your work and ours may perhaps be indicated by the fact that you address your letter to President Van Hise, not knowing that his death occurred more than a year and a half ago." Birge's victims almost never responded.

—George Clark Sellery, *Edward Asahel Birge a Memoir* with an "Appraisal of Birge the Limnologist" by C. H. Mortimer, University of Wisconsin Press, Madison, Wisconsin, 1956, pp. 136-137.

### 2. Chancey Juday—A Birge Associate *(from page 6, 18 and 175)*

Juday's first assignment as survey biologist was to study diel migration of zooplankton in Mendota and other lakes of southeastern Wisconsin. After only a year on the survey however, Juday developed tuberculosis and had to leave the Midwest. For the next few years, while he served on the biology or zoology staffs of the Universities of Colorado and California, there was a hiatus in limnology at the University of Wisconsin.

>—Annamarie Beckel, contributing editor, "Breaking New Waters", *Transactions*, The Wisconsin Academy of Sciences, Arts and Letters, Madison, Wisconsin, 1987 edition, pp. 4. Comments attributed to Doctors Frey and Noland.

Juday had to work as long as he could because Birge had arranged for him to receive retirement pay from the Geological and Natural History Survey. Mrs. Juday, after her husband's death expressed her displeasure at Birge's neglect by not consulting on the disposal of Juday's library. She sent his work to the Academy of Natural Sciences in Philadelphia.

>—Annamarie Beckel, contributing editor, "Breaking New Waters", *Transactions*, The Wisconsin Academy of Sciences, Arts and Letters, Madison, Wisconsin, 1987 edition, pp. 94.

### 3. Academic Criteria—Early 1900s *(from pages 6, 104)*

*Women educators in the patriarchal society of the 19th century were beleaguered by many injustices. Requisites for teaching in the high school level required women to submit scholarship and fitness qualifications before the Committee on Examinations and Appointments. They had to provide a diploma from a state college accredited by the Board of Superintendent of Schools.*

As head of the Biology and Physiology Departments at Milwaukee East and South High Schools (1890-1899), Miss Merrill's salary was $1,300 the first year and $100 additional each year of teaching thereafter until a maximum of $1,700 was reached.

Although Miss Merrill had attained that amount by 1894, she continued to teach five more years with no increase in income.

Male heads of departments of Milwaukee school systems were paid approximately $400 more per annum than their female counterparts. As an assistant professor, Miss Merrill accrued greater income from teaching college courses which averaged approximately $400 per year above public school instructors' wages.

Sanford Hooper, principal of Milwaukee East High School in 1897 earned $2,400, the highest attainable wage at that level. In 1910, Merrill was paid $600 to work up results of her South American expedition. Semester teaching fees for the same period average $200 and she was paid $75 for lectures in various departments. *[See ledger cards from the Board of Regents Executive Committee Report on page 104.]*

>—Milwaukee School Board Minutes, 49th Annual Report of School Directors of the City of Milwaukee, June 1908, bound volumes on Annual Reports, Stack 4, State Historical Society of Wisconsin Library.

The Executive Committee of any ward shall have the right at the end of the school year to terminate the service of any woman teacher married, during the term and forfeit her pay from the time of marriage.

>—Milwaukee School Board Minutes, Amendment to Section 2, Article XXIII, September 4, 1884.

Women instructors who were absent or tardy during school sessions had a proportionate part of their salaries deducted with the amount varying from 25 percent with an excuse of illness to full deductions for other causes of absence. There was the interesting case of Miss Lilias Steele who was called before the Committee on Discipline for missing part of one day out of the entire year. Although she had previously worked overtime on several occasions, because she gave no cause for absence, her pay was totally deducted along with a resounding reprimand by the honorable Committee on Discipline. She appealed with no result.

—Milwaukee School Board Minutes, May 1890,
a report from the Honorable Committee on Discipline, pp. 289.

It was voted that women teachers who married, forfeited their teaching contract.

—Milwaukee School Board Minutes, March 8, 1898, from the School Director's Report.

## 4. Birge Earns Ph.D. *(from page 6)*

When the idea of getting the Ph.D. had ripened, Birge went east to Cambridge and consulted Professor Shaler. "I told him," Birge relates in the most finished of his autobiographical notes, "that I thought I knew enough zoology to warrant the degree, but that I was sure that no board of examiners could ask the questions that I could answer. My "advanced work" in zoology was based on Gegenbaur and other German authors, and those older men would never have read them and could not ask questions that I could answer. I shall never forget the feeling which Shaler's reply inspired in me. Quite apart from any personal kindness toward me, he fairly "jumped at" the chance of having a Harvard Ph.D. in a progressive midwestern university. So I went back to Madison, made the application for an examination, etc. When I returned to Harvard for the examination and reported to Shaler, he gave me a most important piece of advice. "Don't say that you "don't know" if the examiners ask you a question that you can't answer. Just make a shot at the answer if you can and then give them some Gegenbaur on an allied subject. They never read Gegenbaur and probably never heard of him. So don't quote him, but give them your version of some of his stuff, mixed with your own work on lakes." I may quote this advice now …Anyway I followed it; the committee recommended the doctorate for me and Harvard granted it." Ph.D., 1878.

—George Clark Sellery, *Edward Asahel Birge a Memoir* with
"An Appraisal of Birge the Limnologist" by C. H. Mortimer,
University of Wisconsin Press, Madison, Wisconsin, 1956, pp. 14-15.

## 5. Student Housing and Greek Letter Societies *(from page 8)*

The mere existence of the Greek-Letter societies provoked criticism. As early as 1888 the Board of Visitors had urged that steps be taken to discourage them, insisting that it was against the best interest of the university students to be members of secret societies. The Visitors claimed that their influence upon their members was on the whole pernicious and against the highest development of the student as an American youth.

—Merle Curti and Vernon Carstensen, *University of Wisconsin—A History, Vol. 1,*
University of Wisconsin Press, Madison, Wisconsin, 1949, pp. 665.
First published in University of Wisconsin Regents Biennial Report, 1887-1888.

Fraternities and sororities, which had become numerous and had begun to acquire chapter houses in the 1890's, continued to increase in numbers in the early years of this century. Some began to build their somewhat ostentatious houses. These associations represented different things to different University groups. To their own members, the fraternities offered a place to live, social prestige, a circle of friends, and often a focus of lasting interest; to other students, fraternity row was the gold coast of the University, the fraternity men were the sons of wealth, wasteful of time and money, dexterous in controlling University politics, aloof and unfriendly; to the faculty these associations often represented organizations which, because of the housing and boarding service offered, must be tolerated, regulated, and from time to time so controlled as to protect freshmen from their clutches; to many old alumni they were enduring points of interest in the vast stretches of the growing institution; to the regents, they were sometimes an annoyance, even an embarrassment; and to members of the legislature, they supplied an ever available target for attack because of their putative expensiveness, exclusiveness, lack of democracy and immorality.

Through the years these organizations were the subject of a running argument, animated but of course inconclusive. Van Hise and many of the faculty felt that in an ideal world the University could very well get along without them, but that world had never emerged. Writing to the president of Smith College, Van Hise declared that if a college could provide halls under college supervision, he could "see no possible occasion for Greek-letter fraternities." Thus it was the lack of University housing which conditioned all faculty and regent consideration of the fraternity problem. The legislature in 1907 threatened to abolish fraternities and in 1909 asked the regents to investigate the charges that the organizations were undemocratic. This task the regents passed on to a faculty committee. The committee made a full investigation and brought in a report which probably surprised no one. The committee found that the fraternities were to an extent exclusive, but not extensively antidemocratic, snobbish or clannish. It reported that other than well-to-do students were members. Indeed the committee found that about 27 percent of the fraternity men were partially self-supporting and 7 percent entirely so. The management of the chapter houses was not efficient, but it was improving. The committee found that the fraternity members were definitely deficient in scholarship and since the members came from better preparatory schools than the non-fraternity students, they thought this condition was to be explained largely in terms of the opportunities which existed for wasting time with congenial companions. Over emphasis on social activities contributed to this record. So far not much had been done by the faculty to correct these evils. On the other hand, the fraternities offered social advantages to their members. The moral standards of the members differed little from those of other students. The societies offered much needed housing for students. The only adequate substitute for the fraternities, the committee felt, would be an extensive University dormitory system. Until this was provided, the committee declared, "No action looking toward their abolition would seem justified." In the meantime, the committee recommended, no freshmen should be permitted to join, or live in the fraternity houses, and no student on probation ought to be initiated. The regents accepted the report and used it as the basis of their report to the legislature the next year.

—George Clark Sellery, *Edward Asahel Birge a Memoir* with
"An Appraisal of Birge the Limnologist" by C. H. Mortimer,
University of Wisconsin Press, Madison, Wisconsin, 1956, pp. 500-501.

## 6. University Politics *(from pages 7, 13, 14, 105)*

Dean Birge, in his report to Van Hise in 1908, declared that the teacher-training course offered "educational problems of considerable difficulty." The rapid increase in the number of high schools and the expansion of their work had created a large demand for teachers, but pay was still low. The professional life of a teacher was short. One quarter of the teachers each year were beginners. Young men used teaching as a steppingstone to other professions; young women, said Birge, "naturally and rightly" looked upon teaching as an occupation to fill the period between graduation from the University and marriage. Birge looked forward to a change whereby the teaching profession would become stabilized but believed it would not come soon. In the meantime, the University faced the difficult task of providing courses for those who proposed to enter teaching as a permanent profession and for those who should be in it only temporarily. Professional standards must not be placed so high as to exclude the latter.

It was this temporary group which concerned Birge two years later when again he discussed the matter, pointing out that the average period of service for high school teachers was only three or four years. For the University to provide extensive professional training to people who would remain in the profession for only a few years was wasteful. When only a college degree had been required to teach, not much special training was lost if a person remained in the profession only a few years. But conditions had changed: "The student's work in college is becoming controlled and dominated by the brief period of teaching which may succeed it. It loses much of its value as a liberal course of study, and the short period which the student gives to the profession prevents him from realizing any considerable part of its proper value as a professional course."

In establishing the course for teachers, the faculty had agreed: (1) that professional study should be placed in the later years of the course and much of it, if possible, should be gained in graduate work; (2) that under present conditions it was not practicable to require graduate training as a condition of teaching, so professional training must be provided within the undergraduate courses; (3) special care should be taken to develop the professional spirit in those expecting to teach. Accordingly, the faculty provided both an undergraduate and advanced course for the training of teachers, the first to be offered in the senior year, and the second, consisting of twelve semester hours, to be taken during summer school or one semester of graduate work.

—Merle Curti and Vernon Carstensen, *University of Wisconsin—A History, Vol. II*,
University of Wisconsin Press, Madison, Wisconsin, 1949, pp. 256-257.
First published in University of Wisconsin Regents Biennial Report, 1907-1908, pp.53.

The problem of communication was exacerbated by the lack of seminars or any formal set-up for the exchange of ideas. There were certainly discussions around the poker table in addition to the "I'll raise you 30," as well as conversations that were a combination of shop talk and story-telling around bonfires on the beach. In later years students did discuss ideas and research problems over the dinner table. Neither Birge nor Juday, however, participated in these activities. Nor did Birge and Juday introduce their students to the many European researchers who visited Trout Lake or the university.

—Annamarie Beckel, contributing editor, "Breaking New Waters", *Transactions*,
Wisconsin Academy of Sciences, Arts and Letters, Madison, Wisconsin, 1987 edition, pp. 18.

"In the four summers I was here we didn't have a single seminar presentation. This is one reason why theory and empirical ideas simply did not come out in the open because we did not have this kind of discussion."

—R. Pennak, 1983, "History of Limnology in Wisconsin Conference."

Graduate student theses were, for the most part, independently conceived and executed. There was very little guidance from either Birge or Juday. Students were given a free hand and time to do their own research while they worked for Birge and Juday, but it was an individualistic effort—the student was on his own to sink or swim. Fortunately, most swam.

—Annamarie Beckel, contributing editor, "Breaking New Waters", *Transactions*, Wisconsin Academy of Sciences, Arts and Letters, Madison, Wisconsin, 1987 edition, pp. 18.

By 1908, the Graduate School formalized procedures by permitting a candidate whose thesis had been accepted, to either deposit a number of printed copies at the University Library along with a fee or file a copy with an abstract—certified as suitable for scholarly publication and a deposit of fifty dollars. Failure to publish and deposit (in certain cases, 100 copies) within a given length of time, delayed or forfeited the Ph.D. degree. Post graduate courses leading to Ph. D. degrees remained in debate up to 1900. In 1881, Dean Birge, who headed a committee on the question, indicated the situation was merely perfunctory. The faculty then requested the regents to grant no Ph.D. degrees and Birge, determined not to show favoritism, did not alleviate positions in the department of Zoology until 1912. He was considered to be extremely frugal, puritanical and at times, imperious demeanor. In physical stature he was small and wiry. It was know that Birge was reluctant to increase salaries, an act not commensurate with other department heads. He once told a colleague that, "the only thing a scholar needs is, a table, chair and couch."

—Merle Curti and Vernon Carstensen, University of Wisconsin—A History, Vol. II, University of Wisconsin Press, Madison, Wisconsin, 1949, pp. 368-372. First published in the Graduate Office Board of Regents minutes of the Graduate Committee of the University of Wisconsin.

"Dr. Birge and Bob LaFollette Sr., got to be rather bitter enemies when LaFollette was governor *[Birge was then university president]*. They differed on how they thought the university should be operated. Dr. Birge was quite insistent on trying to keep the activities pretty well within the bounds of the campus, while LaFollette wanted the university spread out, to be of more service to the people of the state. "So when LaFollette's son Phil became governor, that animosity apparently continued because with one stroke of the pen, he wiped out the funds for the natural history part of the budget, which left Juday and me out of a job." But the university came to the rescue and took it over. So after that it was funded by a university appropriation and Juday was given university status [full professorship in zoology, formerly he had been a half-time lecturer]." E. Schneberger, 1983, "History of Limnology in Wisconsin Conference."

—Annamarie Beckel, contributing editor, "Breaking New Waters", *Transactions*, Wisconsin Academy of Sciences, Arts and Letters, Madison, Wisconsin, 1987 edition, pp. 18.

**7. Birge's Private Life** *(from pages 13, 15)*

Dr. Birge's teaching, administration, and research were facilitated—perhaps one might say were made possible—by a satisfying social life, which was built around a happy home. He lived at 744 Langdon Street, 1883-1920, in the house he and his wife had planned; in the president's house at the corner of Park and Langdon, 1920-25; and at 2011 Van Hise Avenue, 1925-50. His wife, Anna Grant Birge, was a true help mate. To her he turned over his salary, to spend and to save; he insisted from the beginning that she should have a maid—and there were very few maids, for they liked their position; his wife insisted that he should take a short nap after lunch, before resuming his work in the afternoon; she acquiesced in his wish that she should go East with the children, Edward and Anna, and visit their folks, every summer, while he was busy with his field work.

—George Clark Sellery, *Edward Asahel Birge a Memoir* with
"An Appraisal of Birge the Limnologist" by C. H. Mortimer,
University of Wisconsin Press, Madison, Wisconsin, 1956, pp. 54-55.

To be a scholar, you do not need to leave your room. Remain sitting at your table and listen, simply wait. Do not even wait, be quite still and solitary. The world will freely offer itself to you to be unmasked, it has no choice, it will roll in ecstacy at your feet.

—George Clark Sellery, *Edward Asahel Birge a Memoir*
with "An Appraisal of Birge the Limnologist" by C. H. Mortimer,
University of Wisconsin Press, Madison, Wisconsin, 1956, pp. 143.
Quote used by Birge, in a sermon to his Sunday school class at the
First Congregational Church in Madison, from Franz Kafka's *The Great Wall of China*,
"Reflections", as first documented in the Edward Kremers Papers, State Historical Society of
Wisconsin, Manuscript N. B., Box 4, folder 9-13, 1887, and Box 5, Reminiscences, folder 3.

**8. The Allen Survey** *(from pages 137, 156)*

*Current Opinion* reviewed the press comment on the Allen Survey. Hermon C. Bumpus, formerly university business manager and then president of Tufts College, wrote critically of the Allen Survey in *School Review*, declaring that efficiency cannot be increased through investigations or advice of those who are ignorant of University purposes and out of sympathy with University methods. Allen attempted to defend himself. A letter to the *Dial* merely provoked further critical comment by two University professors, George C. Comstock and William Ellery Leonard. Allen protested to the *Nation* in a long letter denying the charges made in columns, but he got no sympathy. In September, the *Educational Review* carried an article by John Loomis Sturtevant giving mild support to Allen, but Dean Birge and Professor Jastrow answered it promptly. In December, George H. Mead, professor of philosophy at the University of Chicago, in a long, thoughtful article in the *Survey* reviewed the relation of the University to the state and the Allen report. Allen's survey, Mead declared, was 'an example of how not to survey a university." Again Allen was given space to answer, but he could not meet the objections Mead had raised. Nowhere in the nonpartisan press was Allen's survey approved. At the end of the year Professor Ellwood P. Cubberley of Stanford pronounced it an "outrageous investigation."

—George Clark Sellery, *Edward Asahel Birge a Memoir* with "An Appraisal of Birge the Limnologist" by C. H. Mortimer, University of Wisconsin Press, Madison, Wisconsin, 1956, pp. 280-281.

The total cost of the survey was claimed by Allen to have been $13,200 plus the $2,000 for printing. The secretary of the Board of Public Affairs reported that it had cost $17,802.47 as of July 9, 1915, but that not all bills were in. Secretary W. W. Powell to Governor Emanuel L. Phillips, July 9, 1915, Phillipp Papers. Dean Sellery later estimated the cost, including the time the University was required to spend on it, at over $100,000.

—George Clark Sellery, *Edward Asahel Birge a Memoir* with "An Appraisal of Birge the Limnologist" by C. H. Mortimer, University of Wisconsin Press, Madison, Wisconsin, 1956, pp. 281.

*Part of a tea set with a large tray (owned by M. L. Hartridge) made for the British eighteenth century trade and sent on sailing ships to America up to the early nineteenth century from Canton Kilns (usually). Note Christian influence halos on Holy Men.*

## 125TH YEAR SPECIAL ISSUE

## Wisconsin Academy Review
A JOURNAL OF WISCONSIN CULTURE

Wisconsin Academy of Sciences, Arts and Letters

## H.B. Merrill: Early Wisconsin Scientist and Adventurer
*by Merrillyn L. Hartridge*

*Oh, I'd love to roll to Rio*
*Some day before I'm old*
Rudyard Kipling

In 1900 Harriet B. Merrill joined the faculty at the University of Wisconsin as an assistant professor reporting to Dr. Edward Asahel Birge, chair of the university's Department of Zoology. Her experience as head of the science department at Milwaukee Downer College and the years of teaching in Milwaukee high schools had prepared her for this role. She found that her studies in microbiology required more extensive monitoring of a cladoceran in the family Macrothricidae, a study that Birge himself had begun but found little time to research. She had written a monograph on the systematics and anatomy of the genus Daphnia; at the time it was published, she could not have known that it would become a life-long pursuit.

*Harriet Bell Merrill (1863–1915), early Wisconsin scientist, was the first woman to hold the position of vice-president on the Wisconsin Academy council. This photo was taken in 1890 when Merrill was twenty-seven, the year she graduated from the University of Wisconsin with a master's degree. Courtesy State Historical Society of Wisconsin.*

This also was the year that Birge was appointed acting president of the University of Wisconsin, distracting him further from his research. He had begun teaching at the university in January 1876 and was well known among his colleagues. Descriptions of him stress his rather imperious demeanor and disposition. Colleagues remembered his aloof and often brusque manner–he was both disliked and admired. I could find no indication that he gave credit to Harriet Merrill's eventual contributions to science, nor did he appear to give much credit to other male researchers who followed later. In 1902 Merrill had launched a campaign in Madison and Milwaukee to support Birge as candidate for president of the university; in 1903 Charles Van Hise was unanimously elected to the position.

Merrill's early involvement with Birge–their association and correspondence spanned more than twenty-five years–placed her in the center of science activities both in Madison and in Milwaukee. Birge had used his position as dean of the School of Letters and Science to channel funds to the new limnology program on campus. During these years he was active in the Wisconsin Academy of Sciences, Arts and Letters and director of the Wisconsin Geological and Natural History Survey, and he hoped to use these positions to lure scientists to the study of limnology at Madison. Merrill's active role as his research assistant enabled him to further this work. Her research eventually resulted in an authoritative account on the taxonomy of Cladocera.

16

Spring 95 • Wisconsin Academy Review

# Index

## A

Abrams, Bernard, 9
Abreu, Luiz, 31
Academy of Natural Sciences in Philadelphia, 175, 193
Adams, Charles K., University President, 2, 4, 14, 15, 105
Adirondacks, 52
Aid Alliance, 10
alfalfa, 89
Alfonso, Professor Pedro, 82, 83
Algonquin Pequot tribe, 108
"All a scholar needs is one small room with a table, chair and a couch" (Birge), 2
Allen, Charles H., Professor, xiii, 137, 157
Allen investigation, 136
Allen matter, 152
Allen mess, 184
Allen Survey, 137
Allen's Normal School, xiii
Allis-Chalmers, a heavy machinery company of Milwaukee, Wisconsin, 89
alpaca, 95
Alto Sierra, 36
Alves, Francisco da Paulo Rodriguez, 84
Amazon, 39, 56
American Army Corps of Engineers, 85
American church, 88
American Express company, 46
American Microscopial Society, 138
Ames, Mr., 48
Andover, 110
Anopheles, 50
Argentina, 49, 51, 93, 97, 98, 100, 101
Argentine beef, 62
Argentine meat, 89
Argentine Methodist, 57
Argentine minister to the United States, 48
Armadillo, 42
*Art of Fiction*, 104
Ashland, 162
Asunción, 41, 48, 49, 51, 53, 54, 57, 79, 94, 99
*Atlantic Monthly*, 127
Avenue of Royal Palms, 30

## B

Babcock, Dr. Steven M., 34, 81, 115, 140
Baden-Baden, 122
Bahia, 25, 28
balsa disks, 42
Barbados, 30, 69, 85
Barbados, shutter covered house, 69
Barber, Judge, Chicago, 129
Bardeen, Charles, 147
barley, 89
Bascom, 106
Bascom Hill, 103
Bay Rum, popular shaving lotion up through early 1920s or later, 36
Bayfield, 130

Beardsley, Aubrey, popular British graphic artist of the late 1890s whose Eastern styles were followed by avant-garde of the time, 134, 139
Beaver Dam, 126
beef and leather exports, 61
beef and leather industry, 62
*Beginning of the Armadillos*, 19, 21
Belém, 85
Berner, Dr. Dorothy, Temple University, Philadelphia, vii
Bertoni, Señor, 99
Bethel, 110
Bicknell, Mr., United States Department of Agriculture, 37, 38
Bicknell, Mr., (a company legal counsel), 38
Bierbach, Arno, 121
Bierbach, Norton, 121
Binford, Mike, Deevey's student assistant, 175
Biology Building, 131
Birge, Charles, 129
Birge, Dr. Edward Asahel, dean of the zoology department, University of Wisconsin, acting president 1901-1903, president 1918-1925, ix, xiv, xvi, 2, 3, 6, 7, 13, 15, 43, 44, 49, 67, 105, 126, 131, 140, 155, 165, 167, 168, 170, 174, 180, 184, 185, 197
Birge Hall, 131
Bloomington, Indiana, 175, 181
bombilla, 81, 82
Botanical Garden, The, 30
Botocudo, 42
Bowdoin Athenaen honor society, 111
Bowdoin College, 109, 111
Bowdoin College Library, vii

Boyd case, 120
Brazil, 30, 31, 36, 39, 49, 51, 58, 100, 101, 175
breweries, 89
British in Venezuela, 86
Brooks, John L., 175
Brunswick, Maine, xiii, 16, 108, 113
Bryan, Colonel, home of, 68
Bryan, Colonel William Page, minister to Brazil, 37, 70, 72
bubonic plague, 83
Buchanan, Mr., ex-minister to Argentina, 37
Buenos Aires, 31, 33, 53, 57, 75, 76, 89, 90, 91, 94, 100, 101
*Bunops*, xvi, 2, 43, 174, 180
*Bunops scutifrons*, 178
Burgoyne, Major General John, 110
Burleigh, Congressman, of Maine, 20
Burnham Hospital, Champaign, 162, 168
Burrows, 117
Byran, Colonel William Page, 33, 37, 70, 72
*Byron*, 25, 33, 34

# C

Cabot, John Sebastian, 56
Calle Florida, 59
Calle Victoria, 114
Cambridge, Massachusetts, 6
Camp Randall, 133
cannibalistic, 41
Carn, Miss, 137
Carpenter, Frank, American teacher, 36
Casa Rosada, 59
Casper, Lynn Ann, art historian and consultant, Milwaukee, vii

Castelli, Señor, Governor of Misiónes, 65
Catalogue of KAO, xviii
*Ceriodaphnia rigaudi*, 47
cerveza, 81
Ceylon, 39
Chaco, 54
Chambira, 42
Champaign, 161, 162
Chapel, Robert T., Archive Assistant, viii
Chicago, 113, 121, 137, 161, 174, 176
Chicago packing houses, 62, 89
Chicago petition, 106
Chicago St. Paul Railroad Company, 114
Chicago University, ix
cholera, 29, 83
Chydoridae, xvi, 176
*chydorids*, xvi
cigars, 79
Cladocera, xvi, 123, 126, 131, 135, 136, 139, 165, 174, 175, 183, 184
Cleveland administration, 48, 72
Coerne, Louis, 136
coffee of Brazil, 36, 89
coffee plantations, 73
coffee, 36, 81
cog rail, 34
cog train, 35
Colescott, Warrington, Emeritus Professor of art, University of Wisconsin, Madison, vii, viii
Columbia, 149
Columbia Hospital, 162
*Column of the Home, The*, 101
Comstock, Professor, 6
Consulates, 70

Cook County Hospital, 121
Coolie, 95
Cordoba, 91
corn, 89
Cornell University, ix, xiv, 37, 45, 140, 174, 176
Corrientes, 41, 51
cotton, 91
cowhides, 61
Cox, "Sunset", 48
Cremona, 121
Cretaceous, 184
crocodiles, 51
Cruz, Dr. Oswaldo, 82, 83, 84
Cruz's mosquito brigade, 85
Curie, Marie, 2
Cuzco, 60

# D

daguerreotype, 116, 117
Dallman, Dr. John, curator, University of Wisconsin Zoological Museum, vii
Damrosch, Walter, 52
*Daphnia carina*, 142
Daughters of Santos Dumont, 74
Daughters of the American Revolution, xv, 101, 108
de Bourel, Señora Catalina Allen, 101
de Francia, Jose Gaspar, 58
de Sala, Mrs. Alvina V. P., 101
*"Dean Birge ends 40 years with the University of Wisconsin"* (Milwaukee Sentinel), 164
debtors, 88
Decatur, 138

Deevey, Dr. E. S., University of Florida, xiv, 175
Delta Kappa Epsilon, 111
Dent, Edward, 62
Department of Botany, 181
Department of Zoology, ix
Derby, Professor, 34
*Description of Recently Discovered Cladocera from New England* by A. A. Doolittle, 182
Dewey Decimal System, 140, 141
Diana, the huntress, 32
*Diaptomus*, 175, 184
*Diaptomus merrilli*, xvi
Dickenson, Emily, 164
Dietz, C. C., 129
diphtheria, 85
diphtheria bacillus, 83, 85
Director General of Health, 84
Director of the Milwaukee Women's Exchange, 122
Dodson, Dr. Stanley I., University of Wisconsin Department of Zoology, vii, 174, 178, 181
Dominicans, 57
Dousman, 131
Dousman Road, 127
Dumont, Santos, 46, 73, 74

# E

East Andover, Maine, 110
East Andover, Massachusetts, 109
Eastman Kodak, 91
"El Ombú", 55
Elmside Boulevard, 119
Emerson, 110
Emmons, 113
Emmons, Anna Comstock, xiii
*Endentata*, 42
epiphytes, 51
Episcopal House, 137
European demand for Argentine beef, 62
*Europhorbiaceae*, 39
Evans, Dr., 128, 129
expatriates, 87

# F

Fair Oaks subdivision, 118
Fern Room, 114
Fidelity and Deposit Company of Maryland, xiii
Finger Lakes of New York, 148, 167
Finlay, Dr. Carlos, Cuba, 84
fires in the third Wisconsin State Capitol, 107
First Congregational Church, 15
Florida Street, 94
foreign diploma, 91
Four hundred thirteen South Baldwin Street, 119
Francis Street Cabal, 14
Franciscans, 57
Freeland, Dr. James, 109
Freeland, Sarah, 109
Freeman, John Charles, Professor, University of Wisconsin, 17, 24, 25, 123
Frey, Dr. David G., Emeritus Professor of Biology, University of Indiana, vii, 174, 175, 176, 177, 178, 181, 182, 183

# G

Gainsborough, 44
Galsworthy, John, 128
gaucho, 55, 62, 90
Gegenbaur, 6
Geological and Natural History Survey, 175, 193
Germans, 89
giant mantis, 50
Goa, 31
Goldschmidt of Munich, 133
Gondelsheim, Germany, 122
Gondwanaland, 183, 184
government control, 86
Governor's Mansion in Misiónes, 64, 65
Graham, Miss Mary, 98
Gran chaco region, 41
Granell, 149
*Grasshopper*, 74
Great Lakes Biological Survey, 6
Greek-letter fraternities, 8
Green Lake, 131
Greenwood, 126
Grenadiers, 67
Grierson, Dr. Cecilia, 100
Guanas, 42
Guarani, 41, 43, 56, 58, 92
Guarani Indians, 39, 56

# H

Haase, Dr., 127, 129
Haessler, Carl, 133, 145
Hague, The, 86
Hammond, Regent, 130
Hartford, Wisconsin, 181
Hartman, Miss, 101
Hartridge, Dr. Theodore Livingston, vii
Hasler, Dr. Arthur Davis, Emeritus Professor, University of Wisconsin Limnology Center, vii, xvi, 174, 175
Hawthorne, Nathaniel, 109, 111
Heidelberg student rebellion, 122
Hess, Conradine Henriette, 116, 117, 119
Hess, Dena, 123
Hess, Elsie Elizabeth, 121, 122
Hess, Florence Lucille, 120, 121, 122
Hess, Heinrich, 122
Hess, Louis, 119, 123
Hess's Corners, 117, 118, 123
*Hevea brasiliensis*, 39
Honorary Fellow of the University of Chicago, ix, xiv
Hooper, Sanford, 11, 193
Hopkins, Johns, 149
Hotel de Paris at Posadas, Argentina, 63, 77
Hotel de Paris garden, 63
houses shingled with shutters, 69
Hudson, William Henry, Scottish naturalist, 55

# I

"*If God did not exist, it would be necessary to invent him*" (Voltaire), 15
Iguazú, 51
Iguazú Falls, 48, 49, 50, 51, 56
Iguazú River, 51
Ihering, Herr Von, 47

Illinois State University, Urbana, Illinois, 166
Incan gold, 60
International Council of Women, 100
Irving, 110

# J

James, Henry, 104
James, President, 134, 150
Jardin Botanico in Rio, 30
Jesuit missionaries, 56
Jesuits, 56, 57, 58, 60, 93
Journal Club, 147, 148, 152, 159
Juday, Chancey, Indiana University, xvi, 6, 13, 18, 131, 138, 139, 167, 174, 183, 197
Juday estate, 175
Jujuy, 38
Jurassic formation, 38

# K

Kant, 5
Kappa Alpha Theta, 7
Kemper, Bishop, 137
King Makers, 14
Kingsley, Dean, University of Illinois science faculty, 134, 148, 149, 162
Kingsley, J. Sterling, 146
Kingsley, Mary, 17
Kipling, Rudyard, 19, 21, 42
Krause, John, 121

# L

La Boca, 31
La Guaira, 86
La Plata, 39, 41, 53

La Plata Museum, 32
lace, 92, 93
lace mantilla, 94
LaFollette, Robert, Governor, 14, 144, 145
Lake Mendota, xviii, 103
Lake Monona, 118
Lake San Bernadino (Ypacarai), 43
Lake Wingra, 4
Lake Ypacary, 57
Lanusse, Governor of Misiónes, 55
Laurea Honorary Society, ix
Leipzig, 6
Lexington, Massachusetts, 110
Linnaean system, named after Karl von Linne who wrote in Latin and went by the name of Carolus Linnaeus. He became professor of Botany at the University of Upsula, 31
linseed, 89
liverworts, 51
London and River Plate bank, 46
Longfellow, Henry Wadsworth, 109, 110, 111
Lopes, Dr. Ernestina A., 100
López, General Francisco Solano, 58, 59, 65, 66
López, Señor Solano, 65, 66, 67
López, Carlos Antonio, 54, 58
Lord, Mr., 48
lotteries, 86
Lowell, 110
Lowell School, 118
Lupton, Carter, vii
Lynch, Mme. Eliza, 58

# M

Macrothricidae, xvi, 47, 176, 178
*Macrothricids*, xvi
*Macrothrix*, 47
Madison, 115, 122, 132, 143, 170, 177, 181
*Magdalena*, 21, 44, 45, 46
*Maine*, 11
Majestic Building, Milwaukee, Wisconsin, xiii, 114
Majolica, 69
malaria, 29, 66, 83
Malaya, 39
Manaus, Brazil, 39, 40
Manguinhos, 83
manioc, 42
Männechor, 122
Masonic Orders, 57
Massing, Honorable Francis, 121
Mattoccas, 41
Mbayas, 42
McGovern, Francis, 144, 145, 152
McGuire, Dr., 34
McKinley, President, 11, 13, 72
McLaughlin, Dr., 88
measles, 85
Mendoza, 38
Merrill Building, 114
Merrill, Ezekiel, 110
Merrill, Harriet Bell, 174
Merrill, Marie (Mrs. Nathan Emmons), 116, 122
Merrill, Mrs. Silas, 11
Merrill, Roger, 168
Merrill, Samuel, xiii
Merrill, Sarah, 113
Merrill, Uncle Captain Leonard Parker, 109, 111
Merrill's paper, 180
*Merrilli*, 175, 184
Metropolitan, 122
Mexico, 101
Middleton, Dr. W. S., 147
Mihanovich, Nicholas, 53
Mills, 136
Milwaukee, 10, 72, 132, 162, 170
Milwaukee alumna, 105
Milwaukee alumni, 106
Milwaukee bridge engineers, 121
Milwaukee Club, 114, 130
Milwaukee-Downer College, ix, 4, 5, 178
Milwaukee-Downer College catalog, 5
Milwaukee-Downer Science Department, xiv
Milwaukee East High School, ix
*Milwaukee Free Press*, 168
Milwaukee Museum of Natural History, 178
Milwaukee Public Museum, vii, xvii, 17, 25, 41, 50, 52, 92, 108, 109, 140, 141
*Milwaukee Sentinel*, 17, 24, 25, 29, 30, 34, 37, 38, 39, 40, 43, 44, 46, 51, 52, 53, 54, 56, 59, 62, 64, 67, 74, 79, 81, 86, 88, 89, 92, 94, 98
Milwaukee South High School, ix
Milwaukee State Normal School, 9
minister of public instruction, 100
Minister Lord at Buenos Aires, 72
Misiónes, 56, 91
missionaries, 88
Mississippi riverboats, 53
Mitchell, James, 35

*Moina rectirostis*, 142
Monona, Olivia, 122
Monroe Doctrine, 86
Montevideo, 32, 57
Moore, Professor, 133
Morgan, Lloyd, 35
Morgan, Pierpont, 44
mosquito brigades, 84
mosquitos, 50
mosses, 51
Mössner, Elizabeth, 122
Mössner, Dr. Richard, 121
motuca fly, 41
Municipal Bureau at New York University, 157

# N

nanduti, 92, 93
Napoleon I in Europe, 58, 65
National Conscription, xiii
National Historical Building, 132
Nattalock, 109
Natural History Building, 139, 140
*Naturalist on La Plata, The*, 55
Nemesis, the goddess of vengeance, 32
Nesselrode, Count, 45
Nesselrode pudding, 45
New York Finger Lakes, 156, 158
New York Finger Lakes paper, 144, 148, 151
New York lakes, 150
New York Life, 91
New York Life Insurance Company, 91
Newcombe, Dr., 161
Nietzsche, Friedrich, 161
Nineteenth Amendment, August 18, 1920, xiv

Normal School in San Juan, 101
Northwestern Medical School, 121
Nunnemacher, Rudolph, 110

# O

objects of curious disposition, 75
Oconomowoc, 126, 127, 131, 135, 161, 174
Odgen, Dr. Henry Vining, 126, 128, 129, 174
old stagecoach inn, 127
Ombu, 55
Open Door in China, 86
Otto, Adelaide, 116
Otto, Johann, 115, 116
Otto, Johanne, 116
Otto, Wilhelm J., 115
*Outlook*, 128
Owen, Professor, 8, 55
oxcarts, 60, 79

# P

pampas, 55, 62
Pan-American Exhibit, 13
Pan-American Exposition, 11
Panama Canal, 85
Pará, 33, 85
Paraguay, 39, 41, 51, 53, 56, 58, 66, 88, 91, 98
Paraguay River, 53, 54
Paraguayan tea, 91
Paraguayans, 79
Paraná, 39, 53, 56
Paraná River, 41, 49, 51, 53
parasites, 51
Pasteur, 2

Pasteur Institute in Paris, 83
Patagonia, 56
Pavlov, Ivan, 7
Payne, 5
Payne, Mr., 45
Peck, Mr. Edward, 79, 80
pedagogic congress, 100
Pedro, Emperor Dom, II, 31
Pedro, Prince Regent, 30
*Peppys' Diary*, 127, 129
Pepys, Samuel, 44, 45
Pernambuco, 25, 29, 83
Petropolis, 34, 70
Peucinian Society, 111
Pfister Hotel, 72, 114
Phi Beta Kappa, 111
Philippines, 86
Phoenix Hotel, 48, 94
Piraiba, 43
Pirarucu, 43
Pitch Lake, 38
plague, 29, 83
Plaza de Mayo, 57, 67
Plaza Victoria, 59, 60
plazas, 60
Poe, 110
Polish history, 9
polo, 90
poncho, 90, 94, 95
Pope Leo XIII, 72
Portugal, 30
Portuguese naval officer, 31
Prince Regent, 30
Princeton, 138
Protestant mission, 57

# Q

Quarantine flags, 29
Queen Maria, 30

# R

rain forest, 52
rain forest dwellers, 41
Ravdin, Susan, vii
Read, Dr. Allen Walker, New York, Rhodes Scholar, Professor Emeritus of English at Columbia University, viii
Reed, Walter, Virginian, 2, 84
Reid, Dr. J. W., vii
Réjane, the famous French actress, 44
Remedial institutions for the unmanageable delinquents, 10
Republic of Panama, 66
review of the New York lakes, 145
Ricker, 136
Río, 30, 31, 41, 82, 84, 94
Río de la Plata, 31, 41, 81
Rio Botanical Gardens, 166
Rio de Janeiro, 30, 33
Rio to São Paulo run, 34
Ritter, California, 144
Rivero, Edwardo Goncalves, 40
roll of hard candies with holes in the middle (the first Life Saver), 123
Rolled to Rio, 166
Roman Catholic cathedrals, 59
Roosevelt articles, 128
Roosevelt, President Theodore, 66, 86, 119
Rosario, 41
Roux, Pierre Emile, 83

Royal Palm Avenue, 32
rubber companies, 39

# S

*S. S. Byron*, 25
Sabin, Ellen S., 5
Sabin, Ethel, 142, 145, 157
San Bernadino, 57
Santa Barbara, 87
Santos, 33, 34, 46, 83
São Paulo, 33, 35, 46, 47, 73, 74, 81, 87, 98
Sarmiento, President, Argentine educator, 57, 98
Sars, a researcher and scientist who collected Macrothricidae species in the same field of study as Merrill, 47, 142
*Scapholeberis*, 135
scarlet fever, 85
Schermetzler, Mr. Bernard, University of Wisconsin Memorial Library Archives, vii
*Scribner's*, 128
*scutifrons*, 180
Seeger, Consul General, 34, 37, 45, 70
*Selaginella*, 51
Semmelweis, 2
Sewell, Mrs. May Wright, 100, 101
Shaler, Dr., 6
Sheedy, Pat, notorious gambler, 44
Shelford, Victor Ernest, 146, 155
Shutter covered house, Barbados, 69
Siefert, H., 7
Siegfried, 52
Sieker, William C., 123
Silurian deposits, 38

Six hundred thirty Langdon Street, 8
Slaughter, Professor, 14
Slichter, Professor, 14
Smallpox, 83
Smith, Frank, 146, 149, 167
Smithsonian Museum of Natural History, Washington, DC, vii
Socialism at Oxford, 133
Southern Cross constellation, 26
Spencer, 5
Springfield, 138
*Squamosa*, 47
St. John's Lutheran Church, 115
St. John's Military Academy, 137
St. Louis Exposition, 34
St. Louis Fair, 37
St. Louis Fair committee, 37
St. Mary's Hospital, 128
Standard Oil Company of South America, 37, 38, 45, 48
State Historical Society of Wisconsin, viii, 103, 176
*Stegomyia*, 84
Sterling, Dean, xiii
Stevens Point, Wisconsin, xiii, 161, 162
Stevenson paper, 145, 146, 149, 150, 160
Stevenson, Robert Louis, 35
Stradivari, Antonio, 121
streetcar lines, 76
*Studies of the Physical and Chemical Characteristics of Inland Lakes*, 147
Sullivan, Miss (a Wellesley girl), 48, 51, 62, 63
Sullivan, Theodore G., 34, 37, 45, 48, 62
Systematics and Anatomy of the Genus *Daphnia, The*, xiv

# T

Teatro Amazonas, 40
theological schools, 58
Thetas, 142
*Titanus* beetle, 25
Tobas, 41
*Transactions of the Wisconsin Academy of Sciences, Arts and Letters*, xv, 138, 175
tribes, 42
tributaries of the Amazon, 53
Trout Lake Research Station, Boulder Junction, ix
tuberculosis, 85
Turner, Frederick Jackson, vice president of Letters and Sciences, ix, 15
Turner, Professor, 14
typhus, 85

# U

Unitarian Church, Milwaukee, 166
University in Heidelberg, 121
University of Brazil, 66
University of Illinois, 180
University of Illinois Archives, Urbana-Champaign, viii
University of Indiana, 131
University of Wisconsin, 122, 123, 131, 136, 137, 151, 152, 174, 178, 181
University of Wisconsin Memorial Library, 168
University of Wisconsin-Madison Center for Limnology, vii, ix
Urbana, Illinois, 113, 130, 132, 146, 161,
Urbana Zoology Department, 6
Uruguay, 58

# V

Van Cleave, Harley Jones, 145, 146, 152, 159
Van Hise, C. P., 8, 14, 18, 142, 145, 155
Venezuela, 86
vice president on the council of Sciences, Arts and Letters of the Wisconsin Academy, ix
vicuna, 95
Vilas, 106, 134, 149
Vilas, Henry, Senator, 14
Vilas, Judge Levi, 122
Vilas, Mr., 55
Voltaire, 15

# W

Ward, Henry Baldwin, viii, 6, 108, 131, 141, 143, 145, 146, 155, 162, 182
Wellesley, 34
Wellesley girl, 48
wheat, 89
Wild, Thadeus, 9
Wilde, John, viii
Wilson, Ellen, 138
Wilson, President T. Woodrow, 137, 138
*Wisconsin Academy of Sciences, Arts and Letters, The*, vii, xiii, 14, 105, 175
Wisconsin Alumni Association, 14, 105
Wisconsin Chapter of the Natural History Society, ix, 15
Wisconsin Geological and Natural History Survey, 6
Wisconsin Idea, 105
Wisconsin Psi chapter, xviii
Wisconsin School of Music, 117
*Wisconsin State Journal*, 118

Wisconsin University, ix
Woltereck, 135
Women's Relief Corps (Drake Post), 9
Women's School Alliance, 9, 11
Women's Self-Government Association, 14
Women's Suffrage, xiv
Wood, Reverend Dr., 57
Woods Hole Biological Laboratories, ix, xiv
Wright, Dr. Stillman, xvi, 175, 184
Wright, Mrs. Robinson, 101

# Y

Yale Library, 175, 176
Yale University, xiv, 175
yellow fever, 33, 36, 83, 84
Yellow Jacks, 29, 33, 36, 83
yerba-mate, 81, 82
Yersin, Alexandre, 83
Young Women's Christian Association, 100

# Z

Zaballos, Señor, Argentine Minister, 48, 55, 72
Zeleny, Professor Charles, viii, 130, 138, 141, 142, 145, 146, 148, 155, 159, 161, 162
zibeline cloth, 97

# About the Author

About the time that Harriet Bell Merrill's family history began in Maine and Massachusetts, Merrillyn Hartridge's fraternal ancestors left England and settled in Hingham, Massachusetts, in the early 1630s. Merrillyn's maternal German side traveled west to Wisconsin, arriving in Madison in 1848. Her home in Madison is filled with memorabilia from past generations, which includes several Merrill mementoes. They have been kept because they evoke fond memories of those who handed them on to her, and are precious remnants of her heritage and history.

An outdoor person, Lynn (as she has been called since her school days), lived most of her years near the shores of Lake Monona in Madison. From an early age, she was captivated by the mysteries of nature. Lynn often took early morning walks during the summer where she enjoyed sketching and painting at the water's edge. In her 11th year, she was suddenly stricken with life threatening peritonitis as the result of a ruptured appendix. In the 1930s the survival rate from such attacks was diagnosed as near to impossible. Although she was fairly recovered in ten months, it changed her life from then on. Though very young, one thing she learned, as the result of her illness, was that persons could never feel totally confined if they enjoyed writing. That summer she created a little book complete with drawings. Ironically it was about a boy from South America, his donkey and the mysterious "cargo" it carried for his village.

Surrounded with books and watercolors from her mother's art room, Lynn filled her days of recovery by writing and illustrating her own stories. She has not put her pen or brushes away since. During World War II, she majored in Applied Art at the University of Wisconsin, was a member of the Woman's Self-Government

*Merrillyn Hartridge*

*Florence Hess*

*Florence L. Hess and D. John Leigh just prior to his leaving to serve in World War I, parents of Merrillyn Leigh Hartridge.*

*D. John Leigh*

Association and the University of Wisconsin Choir.

It was a time when campus politics were not comparable to the highly volatile 60's confrontations, but did contribute to an uncertainty in career plans among students during the war years.

The family observed stringent rationing; her mother was a Red Cross worker and Block Captain. Her father, an accountant with the state treasury department at the Capitol, recalled how his career was interrupted by World War I, but he returned to the same position and then served as Auditor and City Treasurer of Madison, Wisconsin, for 30 years, until his death.

Politically he was an admirable inspiration to those who knew him for his equity of treatment toward all men. The words of Federal Judge Learned Hand of the U. S. Court of Appeals, 2nd Circuit; whom John had met, had a prominent place on the desk in his office of the old City Hall. Lynn inherited his keen sense of patriotism, believing that those who choose to honor our Constitution should call themselves Americans: first.

Lynn studied art under Frances K. Miller, daughter of Joaquin, poet laureate of the West (Cambridge History of American literature) with extension courses through the Corcoran School of Art and during the war, at Washington University School of Art. After two art-history tours abroad, she taught Art and Art History at Madison Area Technical College; has given many lectures on history and art-related subjects; and has written and produced art for advertising in St. Louis, Salt Lake City and locally. She served as contributing editor for *Madison Select* magazine; has written and illustrated for local and national magazines and newspapers over many years, winning local and state awards for art and writing, much of which has been included in anthologies. She produced two programs for television: one educational program for young people on an ABC television affiliate and an adult interview program on NBC television. She has written and illustrated story letters for children who were hospitalized and home-bound; she was a member of the National Tele Media Council; is a member of the Wisconsin Academy of Sciences, Arts and Letters; has served on several civic boards including The State Historical Society of Wisconsin Friends; the board of the Madison Art Guild; board of the former Madison Art Association, the board of the Madison Symphony Orchestra League, twice served as president of the National League of American Pen Women in both Arts and Letters, Madison branch. She was on the Madison Board of the National Society of the Colonial Dames of America, the Madison Pen and Brush Club; Historic Madison, The American Association for State and Local History and other Wisconsin historical societies. Merrillyn is also listed in *Who's Who of American Women* (sixth edition and subsequent volumes).

Both Merrillyn and her husband, retired surgeon Dr. Theodore Hartridge, have found spiritual renewal in forest management of their woodlands and stream in Iowa County since 1971, for which they received a state soil conservation award in 1995. They enjoy hiking (although not as often today) with their eight grandchildren, most of whom have inherited a love of nature and rewards from the wildlife, history of the land, and knowledge that their forbearers walked the same paths.